Journal Information

Subscriptions: *Mathematical Thinking and Learning: An International Journal* is published quarterly by Lawrence Erlbaum Associates, Inc., 10 Industrial Avenue, Mahwah, NJ 07430–2262. Subscriptions for the 2004 volume are available only on a calendar-year basis.

Order subscriptions through the Journal Subscription Department, Lawrence Erlbaum Associates, Inc., 10 Industrial Avenue, Mahwah, NJ 07430–2262.

Claims: Claims for missing copies cannot be honored beyond 4 months after mailing date. Duplicate copies cannot be sent to replace issues not delivered due to failure to notify publisher of change of address.

Change of Address: Send address changes to the Journal Subscription Department, Lawrence Erlbaum Associates, Inc., 10 Industrial Avenue, Mahwah, NJ 07430–2262.

Permissions: Special requests for permission should be sent to the Permissions Department, Lawrence Erlbaum Associates, Inc., 10 Industrial Avenue, Mahwah, NJ 07430–2262.

Abstracts/Indexes: This journal is abstracted or indexed in *PsycINFO/Psychological Abstracts; Zentralblatt für Mathematik/Mathematics Abstracts; ERIC Clearinghouse for Science, Mathematics, and Environmental Education; Current Index to Journals in Education; EBSCOhost Products; Education Index; Education Abstracts; and Australian Education Index.*

Microform Copies: Microform copies of this journal are available through ProQuest Information and Learning, P.O. Box 1346, Ann Arbor, MI 48106–1346.

First published by 2004 Lawrence Erlbaum Associates, Inc.

This edition published 2011 by Routledge
2 Park Square, Milton Park, Abingdon, Oxon OX14 4RN
52 Vanderbilt Avenue, New York, NY 10017

ISBN 13: 978-0-8058-9544-5 (pbk)

ISSN 1098–6065

MATHEMATICAL THINKING AND LEARNING, 6(2), 81–89
Copyright © 2004, Lawrence Erlbaum Associates, Inc.

Learning Trajectories in Mathematics Education

Douglas H. Clements
Julie Sarama
Graduate School of Education
University of Buffalo, State University of New York

Many successful recent approaches to developing innovative mathematics curricula and to conducting research on learning and teaching mathematics have used the construct of "learning trajectories" as a foundation. However, the developers and authors have interpreted and applied this idea in different ways, leading to a need for discussions of these variations and a search for clarifications and shared meanings. Further, the construct of learning trajectories is less than a decade old, but palpably has many roots in previous theories of learning, teaching, and curriculum. The purpose of this special issue is to present several research perspectives on learning trajectories with the intention of encouraging the broader community to reflect on, better define, adopt, adapt, or challenge the concept. This brief article introduces learning trajectories from our perspective. The other articles provide elaboration, examples, and discussion of the construct. They purposefully are intended to be illustrative, exploratory, and provocative with regard to the learning trajectories construct; they are not a set of verification studies.

BACKGROUND

Overviews of recent research-based mathematics curricula reveal several shared characteristics. These include the following: creating and maintaining connections between research and curriculum development as integrated, interactive, processes; using a broad range of scientific methodologies; maintaining close connec-

---

Requests for reprints should be sent to Douglas H. Clements, Department of Learning & Instruction, University of Buffalo, State University of New York, Graduate School of Education, 505 Baldy Hall (North Campus), Buffalo, NY 14260. E-mail: clements@buffalo.edu

Mathematical Thinking and Learning
An International Journal

Editor

Lyn D. English, *Queensland University of Technology, Australia*

Associate Editors

Brian Greer, *San Diego State University, United States*
Richard Lesh, *Purdue University, United States*
Douglas Clements, *SUNY, Buffalo, United States*

Book Review Editor

Gerald Goldin, *Rutgers University, United States*

Editorial Assistant

Shayne Mahon, *Queensland University of Technology, Australia*

Journal Production Editor

Brian JP Craig, *Lawrence Erlbaum Associates, Inc., United States*

Visit *MTL*'s website at **http://www.fed.qut.edu.au/mtl**

Mathematical Thinking and Learning

An International Journal

Volume 6, Number 2, 2004

Special Issue:
Hypothetical Learning Trajectories

Guest Editors:
Douglas H. Clements and Julie Sarama

Contents

tions between tasks and children's mathematical thinking; and using some version of "learning trajectories" (Clements, 2002).

As one example, a team from a realistic mathematics education (RME) group could begin to design curriculum with an anticipatory thought experiment. They formulate a hypothetical learning trajectory that involves conjectures about both a possible learning route that aims at significant mathematical ideas and a specific means that can be used to support and organize learning along this route. The trajectory is conceived of through a thought experiment in which the historical development of mathematics is used as a heuristic. More recently, children's informal solution strategies have also been considered. The original design is a set of instructional tasks with guidelines suggesting an order for the tasks and the types of thinking and learning in which the students can engage as they participate in the instructional tasks. This original design is often not worked out in detail because tasks are revised extensively during classroom testing. That is, the tasks that are actually used in the classroom are determined on a day-to-day basis considering what was learned from enacting the preceding tasks in the classroom. In this second phase, the educational experiment, this preliminary design is refined in a series of intense cyclic processes of deliberations on instructional tasks, the teacher's role, and the wider classroom culture, as these are constituted in the classroom (Gravemeijer, 1999). In the third phase, a "best-case" instructional sequence is constructed. The goal is to develop and describe a more general description of the hypothetical learning trajectories that emerged in specific classrooms that underlies the instructional sequence and to justify it with both theoretical deliberations and empirical data (Gravemeijer, 1994a, 1994b, 1999). This overarching goal is that the local instructional theory will provide a framework that teachers can use to construe hypothetical learning trajectories that fit their own classroom contexts. See Gravemeijer's article in this special issue for the most recent theoretical description.

Learning trajectories can have significance beyond curriculum development. There is evidence that superior teachers use a related conceptual structure. For example, in one study of a reform-based curriculum, the few teachers that had worthwhile, in-depth discussions saw themselves not as moving through a curriculum, but as helping students move through a progression or range of solution methods (Fuson, Carroll, & Drueck, 2000); that is, simultaneously using and modifying a type of hypothetical learning trajectory.

THE LEARNING TRAJECTORY CONSTRUCT

In his seminal work, Simon (1995) stated that a "hypothetical learning trajectory" included "the learning goal, the learning activities, and the thinking and learning in which the students might engage" (p. 133). The name "hypothetical learning trajectory" reflects its roots in a particular constructivist perspective. That is, although the name emphasizes learning over teaching, Simon's description is clearly intended to characterize an essential aspect of pedagogical thinking (i.e., determine the goal, create tasks connected to children's thinking and learning, etc.).

The nascency and complex nature of learning trajectories has led to a variety of interpretations and applications. For example, in contrast to the traditional RME approach and to Simon's (1995) approach, some only emphasize the developmental progressions of learning (what Simon calls hypothetical learning processes) during the creation of a particular curricular or pedagogical context. We believe that, although studying either psychological developmental progressions or instructional sequences separately can be valid research goals and studies of each can and should inform mathematics education, the power and uniqueness of the learning trajectories construct stems from the inextricable interconnection between these two aspects. For our purposes, then, we conceptualize learning trajectories as descriptions of children's thinking and learning in a specific mathematical domain and a related, conjectured route through a set of instructional tasks designed to engender those mental processes or actions hypothesized to move children through a developmental progression of levels of thinking, created with the intent of supporting children's achievement of specific goals in that mathematical domain (cf. Clements, 2002; Gravemeijer, 1999; Simon, 1995).

Consider the developmental progression aspect and the instructional task aspect in turn. First (from our perspective), one specifies learning models that reflect natural developmental progressions (at least for a given age range of students in a particular culture) identified in theoretically and empirically grounded models of children's thinking, learning, and development (Carpenter & Moser, 1984; Griffin & Case, 1997). That is, researchers build a cognitive model of students' learning that is sufficiently explicit to describe the processes involved in the construction of the goal mathematics across several qualitatively distinct structural levels of increasing sophistication, complexity, abstraction, power, and generality. This constructivist aspect distinguishes the learning trajectory approach from previous instructional design models that, for example, used reductionist techniques to break a goal competence into subskills, based on an adult's perspective. This is illustrated by Fuson's (1997) curriculum, which was based on a model of children's solving of word problems, including moving through increasingly difficult types of word problems based on the model. The theory is that learning consistent with such natural developmental progressions is more effective, efficient, and generative for the student than learning that does not follow these paths.[1] Research indicates that providing knowledge of children's thinking and learning in the targeted

---

[1]However, such progressions are not isolated from cultural influences, especially intentional instruction. For example, we postulate that progress through nongenetic levels of thinking (e.g., in geometry) is determined more by social influences, and specifically instruction, than by age-linked development (Clements, Battista, & Sarama, 2001).

subject-matter domain can substantially affect curriculum design by focusing it on teaching and learning (Tamir, 1988; Walker, 1992). This leads our discussion to the second aspect.

The second aspect of learning trajectories is an instructional sequence. In our work, these are composed of key tasks designed to promote learning at a particular conceptual level or benchmark in the development progression. Extant research is used to identify tasks as effective in promoting the learning of students at each level by encouraging children to construct the concepts and skills that characterize the succeeding level. That is, we hypothesize the specific mental constructions (i.e., mental actions-on-objects) and patterns of thinking that constitute children's thinking at each level. We design tasks that include external objects and actions that mirror the hypothesized mathematical activity of students as closely as possible; for example, objects can be shapes or sticks and actions can be creating, copying, uniting, disembedding, and hiding both individual units and composite units. Tasks require children to apply, externally and mentally, the actions and objects of the goal level of thinking (see Clements & Battista, 2000, for further description and examples of mathematical objects/concepts and mathematical actions/processes that operate on them). These tasks are, of course, sequenced corresponding to the order of the developmental progressions to complete the hypothesized learning trajectory. The main theoretical claim is that such tasks will constitute a particularly efficacious educational program. However, there is no implication that the task sequence is the only, or the best, path for learning and teaching, only that it is hypothesized to be one fecund route. Further, societally determined values and goals are substantive components of any curriculum (Confrey, 1996; Hiebert, 1999; National Research Council, 2002; Tyler, 1949); research cannot ignore or determine these components (cf. Lester & Wiliam, 2002).

Thus, a complete hypothetical learning trajectory includes all three aspects: the learning goal, developmental progressions of thinking and learning, and sequence of instructional tasks. The synergism between the latter two aspects has already been described. Less obvious is that their integration can produce novel results, even within the local theoretical fields of psychology and pedagogy. The enactment of an effective, complete learning trajectory can actually alter developmental progressions or expectations previously established by psychological studies because it opens up new paths for learning and development. This, of course, reflects the traditional, if oversimplified, debate between Vygotsky (1934/1986) and Piaget and Szeminska (1952) regarding the priority of development over learning. We believe that learning trajectory research, along with other research corpi, supports the Vygotskian position that, at least in some domains and some ways, learning and teaching tasks can change the course of development. Such an enactment based on the fine-grain cognitive analysis of the developmental progression and the similarly detailed analysis of the instruction tasks provides a more elaborated theoretical base for curriculum and instruction than is often available and can also open instructional approaches or avenues not previously considered. Note that these simultaneous detailed analyses distinguish the hypothetical learning trajectory construct from other lines of inquiry, even such closely related work as the van Hiele's simultaneous dissertations (van Hiele, 1959/1985; van Hiele-Geldof, 1984).

Creation and use of learning trajectories always implies conceptual analysis: "The actual process of thinking remains invisible and so do the concepts it uses and the raw material of which they are composed" (von Glaserfeld, 1995, p. 77). Further, an overarching research goal in the field of learning trajectories is to generate knowledge of learning and teaching. Therefore, scientific processes (e.g., documenting decisions, rationales, and conditions; hypothesizing mechanisms; predicting events; and checking those predictions) must be carefully followed and recorded. Further, authors of models of teaching and learning should consider a wide variety of individual, social, and contextual aspects (e.g., Bauersfeld, 1980; Cobb, 2001; McClain & Cobb, 2001; Schofield, 1995; Secada, 1992). That is, a concerted effort should be made to view the curriculum, and the teaching and learning process, through multiple conceptual lenses (Schoenfeld, 2002), examining both underlying assumptions and data from as many alternate, often incompatible, perspectives as possible (Lester & Wiliam, 2002).

The construct of hypothetical learning trajectories is a cognitive tool grounded in constructivism. It has been adapted for use with a social perspective (e.g., Cobb, 2001), reconceptualizing the learning trajectory construct as a "sequence (or set) of (taken-as-shared) classroom mathematics practices that emerges through interaction (especially through classroom discourse—with the proactive involvement of the teacher)" (Yackel, personal communication, July 30, 2003). Yackel does not use learning trajectories to describe the learning of individual students; this interpretation differs from ours, which includes analyses at both the group and individual levels. However, it does reveal another application, and thus unique contribution, of the hypothetical learning trajectory construct.

In this description, we have taken the perspective of the researcher or researchers/curriculum developer who often writes for a general audience. It is important to note that learning trajectories could and should be reconceptualized or created by small groups or individual teachers, so they are based on more intimate knowledge of the particular students involved—their extant knowledge, learning preferences, and engagement in certain task types or contexts (Simon and Tzur, this issue). Indeed, a priori learning trajectories are always hypothetical in that the actual learning and teaching, and the teacher's recognition of these, cannot be completely known in advance (Simon, 1995). The teacher must construct new models of children's mathematics as they interact with children around the instructional tasks and thus alter their own knowledge of children and future instructional strategies and paths. Thus, the realized learning trajectory, the taken-as-shared practices and understandings, are emergent.

ARTICLES IN THIS SPECIAL ISSUE

This special issue contains articles by several leaders in inventing, elaborating, or contesting the construct of hypothetical learning trajectories. Simon (1995) introduced the construct hypothetical learning trajectory as a way to describe the pedagogical thinking involved in teaching mathematics for understanding. In their article, Simon and Tzur (this issue) elaborate this formulation within their constructivist framework by describing (a) a mechanism, reflection on task-effect relation, that is an elaboration of Piaget's (2001) reflective abstraction and (b) the ways such a mechanism can structure the use of each of several components of the hypothetical learning trajectory. Their mechanism, which explicates the role of mathematical tasks in mathematics concept development, addresses the learning paradox (Bereiter, 1985; Pascual-Leone, 1976).

Gravemeijer (this issue) describes what his interpretation of hypothetical learning trajectories offers to the reform of mathematics education. He contrasts that interpretation with classical instructional design theories that, in his opinion, do not fit mathematics education that tries to capitalize on the inventions of the students. This contrast is especially useful in that it clarifies some unique characteristics of hypothetical learning trajectories. Further, Gravemeijer illustrates his interpretation with a local instruction theory on addition and subtraction, showing how classroom teachers can use such theories to construe hypothetical learning trajectories fitting their classroom contexts.

Steffe (this issue) emphasizes that, "through the construction of learning trajectories that are co-produced by children, it is possible to construct learning trajectories of children that include an account of one's own ways and means of acting and operating as a teacher." He abstracts learning trajectories from acts of teaching children and argues that this is of critical importance, illustrating the process with research on children's construction of partitioning schemes and commensurate fractions.

Clements, Wilson, and Sarama (this issue) describe the genesis of a hypothetical learning trajectory involving young children's composition of geometric shape and present a multimethod evaluation of the developmental progression underlying it (however, instructional sequences were always a component of the research). Clements and Sarama's complete approach is to consider research on both aspects—learning and teaching for the initial hypothetical learning trajectory—and then design, iteratively test, and recognize that individual differences, variations, alternate routes, and further refinements are not only a goal, but a continual requirement (Clements, 2002; Clements & Battista, 2000; Sarama & Clements, in press). They are presently conducting research on the complete hypothetical learning trajectory.

Battista (this issue) argues that, to be scientific, assessment must be linked to research on student learning and cognition. He provides an illustration in the framework of his work on cognition-based assessment, an assessment system to detail the cognitive underpinnings of the progress students make in developing understanding of specific mathematics topics. This, then, also emphasizes the developmental progression aspect only, although the ideas were formed in the context of curriculum development and evaluation (similar to the work of Clements, Wilson, and Sarama, this issue).

Lesh and Yoon (this issue) present a challenge to an overly narrow view of learning trajectories as inflexible, narrow channels of learning (a view we do not believe is held by authors of the other articles, but a possible misinterpretation of the construct) in emphasizing a learning terrain that the authors visualize as an inverted genetic inheritance tree, in which great-grandchildren can trace their evolution from multiple lineages. From our view, Lesh and Yoon illustrate important issues, but we believe the differences can be at least in part an issue of grain size. That is, students who create various models can be following different but equally important learning trajectories (in the small); teachers would benefit from assessing these. More important, especially viewing learning trajectories as occurring in a social setting, each model-eliciting activity can (and probably should) be placed in larger cycles within a broad hypothetical learning trajectory that describes the teachers' and researchers' mathematical goals.

Baroody, Cibulskis, Lai, and Li (this issue) discuss all these articles in their reaction. Baroody and his colleagues provide a useful comparison to previous approaches, comments and critiques of all the other articles, and suggestions for extending the various approaches.

FINAL WORDS

We believe that the notion of hypothetical learning trajectories is a unique and substantive contribution to the field. The construct differs from other models in that it involves self-reflexive constructivism and includes the simultaneous consideration of mathematics goals, models of children's thinking, teachers' and researchers' models of children's thinking, sequences of instructional tasks, and the interaction of these at a detailed level of analyses of processes.

ACKNOWLEDGMENT

This article was supported in part by the National Science Foundation under Grants ESI–9730804, "Building Blocks—Foundations for Mathematical Thinking, Pre-Kindergarten to Grade 2: Research-based Materials Development" (Douglas H. Clements and Julie Sarama, Co-PIs) and REC–9903409, "Technology-Enhanced Learning of Geometry in Elementary Schools" (Daniel Watt, Douglas H. Clements, and Richard Lehrer, Co-PIs), and by the Interagency Educational Research Initiative (NSF, DOE, and NICHHD) Grant REC–0228440,

"Scaling Up the Implementation of a Pre-Kindergarten Mathematics Curricula: Teaching for Understanding with Trajectories and Technologies" (D. H. Clements, J. Sarama, A. Klein, & P. Starkey, co-PIs). Any opinions, findings, and conclusions or recommendations expressed in this material are those of the author and do not necessarily reflect the views of the National Science Foundation.

REFERENCES

Bauersfeld, H. (1980). Hidden dimensions in the so-called reality of a mathematics classroom. *Educational Studies in Mathematics, 11,* 23–41.

Bereiter, C. (1985). Toward a solution of the learning paradox. *Review of Educational Research, 55,* 201–226.

Carpenter, T. P., & Moser, J. M. (1984). The acquisition of addition and subtraction concepts in grades one through three. *Journal for Research in Mathematics Education, 15,* 179–202.

Clements, D. H. (2002). Linking research and curriculum development. In L. D. English (Ed.), *Handbook of international research in mathematics education* (pp. 599–630). Mahwah, NJ: Lawrence Erlbaum Associates, Inc.

Clements, D. H., & Battista, M. T. (2000). Designing effective software. In A. E. Kelly & R. A. Lesh (Eds.), *Handbook of research design in mathematics and science education* (pp. 761–776). Mahwah, NJ: Lawrence Erlbaum Associates, Inc.

Clements, D. H., Battista, M. T., & Sarama, J. (2001). Logo and geometry. *Journal for Research in Mathematics Education Monograph Series, 10.* Reston, VA: National Council of Teachers of Mathematics.

Cobb, P. (2001). Supporting the improvement of learning and teaching in social and institutional context. In S. Carver & D. Klahr (Eds.), *Cognition and instruction: Twenty-five years of progress* (pp. 455–478). Mahwah, NJ: Lawrence Erlbaum Associates, Inc.

Confrey, J. (1996). The role of new technologies in designing mathematics education. In C. Fisher, D. C. Dwyer, & K. Yocam (Eds.), *Education and technology, reflections on computing in the classroom* (pp. 129–149). San Francisco: Apple Press.

Fuson, K. C. (1997). Research-based mathematics curricula: New educational goals require programs of four interacting levels of research. *Issues in Education, 3,* 67–79.

Fuson, K. C., Carroll, W. M., & Drueck, J. V. (2000). Achievement results for second and third graders using the Standards-based curriculum *Everyday Mathematics. Journal for Research in Mathematics Education, 31,* 277–295.

Gravemeijer, K. P. E. (1994a). *Developing realistic mathematics instruction.* Utrecht, The Netherlands: Freudenthal Institute.

Gravemeijer, K. P. E. (1994b). Educational development and developmental research in mathematics education. *Journal for Research in Mathematics Education, 25,* 443–471.

Gravemeijer, K. P. E. (1999). How emergent models may foster the constitution of formal mathematics. *Mathematical Thinking and Learning, 1,* 155–177.

Griffin, S., & Case, R. (1997). Re-thinking the primary school math curriculum: An approach based on cognitive science. *Issues in Education, 3,* 1–49.

Hiebert, J. C. (1999). Relationships between research and the NCTM Standards. *Journal for Research in Mathematics Education, 30,* 3–19.

Lester, F. K., Jr., & Wiliam, D. (2002). On the purpose of mathematics education research: Making productive contributions to policy and practice. In L. D. English (Ed.), *Handbook of International Research in Mathematics Education* (pp. 489–506). Mahwah, NJ: Lawrence Erlbaum Associates, Inc.

McClain, K., & Cobb, P. (2001). An analysis of development of sociomathematical norms in one first-grade classroom. *Journal for Research in Mathematics Education, 32,* 236–266.

National Research Council. (2002). Scientific research in education. In R. J. Shavelson & L. Towne (Eds.). Washington, DC: National Academy Press.

Pascual-Leone, J. (1976). A view of cognition from a formalist's perspective. In K. F. R. J. A. Meacham (Ed.), *The developing individual in a changing world: Vol. I Historical and cultural issues*. The Hague, The Netherlands: Mouton.

Piaget, J. (2001). *Studies in reflecting abstraction*. Sussex, England: Psychology Press.

Piaget, J., & Szeminska, A. (1952). *The child's conception of number*. London: Routledge and Kegan Paul.

Sarama, J., & Clements, D. H. (in press). Linking research and software development. In K. Heid & G. Blume (Eds.), *Technology in the learning and teaching of mathematics: Syntheses and perspectives*. New York: Information Age Publishing.

Schoenfeld, A. H. (2002). Research methods in (mathematics) education. In L. D. English (Ed.), *Handbook of international research in mathematics education* (pp. 435–487). Mahwah, NJ: Lawrence Erlbaum Associates, Inc.

Schofield, J. W. (1995). *Computers and classroom culture*. Cambridge, MA: Cambridge University Press.

Secada, W. G. (1992). Race, ethnicity, social class, language, and achievement in mathematics. In D. A. Grouws (Ed.), *Handbook of research on mathematics teaching and learning* (pp. 623–660). New York: Macmillan.

Simon, M. A. (1995). Reconstructing mathematics pedagogy from a constructivist perspective. *Journal for Research in Mathematics Education, 26*, 114–145.

Tamir, P. (1988). The role of pre-planning curriculum evaluation in science education. *Journal of Curriculum Studies, 20*, 257–262.

Tyler, R. W. (1949). *Basic principles of curriculum and instruction*. Chicago: University of Chicago Press.

van Hiele, P. M. (1985). The child's thought and geometry. In D. Fuys, D. Geddes, & R. Tischler (Eds.), *English translation of selected writings of Dina van Hiele-Geldof and Pierre M. van Hiele* (pp. 243–252). Brooklyn, NY: Brooklyn College, School of Education. (Original work published 1959) (ERIC Document Reproduction Service No. 289 697)

van Hiele-Geldof, D. (1984). The didactics of geometry in the lowest class of secondary school (M. Verdonck, Trans.). In D. Fuys, D. Geddes, & R. Tischler (Eds.), *English translation of selected writings of Dina van Hiele-Geldof and Pierre M. van Hiele* (pp. 1–214). Brooklyn, NY: Brooklyn College, School of Education. (ERIC Document Reproduction Service No. 289 697).

von Glaserfeld, E. (1995). *Radical constructivism: A way of knowing and learning*. London: The Falmer.

Vygotsky, L. S. (1984). *Thought and language*. Cambridge, MA: MIT Press. (Original work published 1934)

Walker, D. F. (1992). Methodological issues in curriculum research. In P. W. Jackson (Ed.), *Handbook of research on curriculum* (pp. 98–118). New York: Macmillan.

MATHEMATICAL THINKING AND LEARNING, 6(2), 91–104

Explicating the Role of Mathematical Tasks in Conceptual Learning: An Elaboration of the Hypothetical Learning Trajectory

Martin A. Simon

Department of Curriculum and Instruction
Penn State University

Ron Tzur

Department of Mathematics, Science, and Technology Education
North Carolina State University

Simon's (1995) development of the construct of *hypothetical learning trajectory* (HLT) offered a description of key aspects of planning mathematics lessons. An HLT consists of the goal for the students' learning, the mathematical tasks that will be used to promote student learning, and hypotheses about the process of the students' learning. However, the construct of HLT provided no framework for thinking about the learning process, the selection of mathematical task, or the role of the mathematical tasks in the learning process. Such a framework could contribute significantly to the generation of useful HLTs. In this article we demonstrate how an elaboration of reflective abstraction (i.e., reflection on activity-effect relationships), postulated by Simon, Tzur, Heinz, and Kinzel (in press), can provide such a framework and thus a theoretical elaboration of the HLT construct.

In recent efforts to improve the quality of mathematics education in the United States, considerable attention has been focused on mathematical tasks. The National Council of Teachers of Mathematics (2000) considered mathematical tasks to be key to the learning of important mathematics. Lappan and Briars (1995) asserted that selecting activities or tasks is the most significant decision affecting student learning. Krainer (1993) affirmed, "Powerful tasks are important points of contact between the actions of the teacher and those of the student" (p. 68). If

Requests for reprints should be sent to Martin A. Simon, Penn State University, Department of Curriculum and Instruction, Chambers 266, University Park, PA 16802. E-mail: msimon@psu.edu

mathematical tasks play such a key role in the effectiveness of mathematics instruction, how do we think about the selection of tasks?

One approach involves the selection of "high-level cognitively complex tasks [that can promote] the capacity to think, reason, and problem solve" (Smith & Stein, 1998, p. 344). This emphasis is based on the idea that if students are challenged at an appropriate level with nonroutine tasks, they develop their cognitive abilities and engage in rich mathematical conversations. Indeed if more time were spent in classrooms with students engaged in working on cognitively demanding nonroutine tasks, as opposed to exercises in which a known procedure is practiced, students' opportunities for thinking and learning would likely be enhanced. Another common approach is to select learning tasks that encourage engagement with the concept to be learned (c.f., Bell, 1993; van Boxtel, van der Linden, & Kanselaar, 2000). This approach can also provide opportunities for student learning. Prior research has contributed ways of characterizing mathematical tasks (c.f., Goldin & McClintock, 1985; Krainer, 1993). However, these contributions leave as implicit the relationship of the tasks to the students' learning processes.

Ainley and Pratt (2002) characterized a planning paradox that mathematics educators confront:

> If teachers plan from objectives, the tasks they set are likely to be unrewarding for the children and mathematically impoverished. But if teachers plan from tasks, the children's activity is likely to be unfocussed and learning difficult to assess. (p. 18)

Ainley and Pratt proposed purpose and utility as features of task design that resolve the planning paradox. We argue that the framework we present in the following discussion affords attention to students' learning processes and contributes significantly to avoiding this paradox.

One of the significant problems facing mathematics education is how to promote students' development of new mathematical concepts (e.g., ratio, derivative, variation), particularly those whose development is often unsure (Bereiter, 1985). To meet this challenge, mathematics educators need an understanding of learning processes and the role of mathematical tasks in the learning process. It is toward this end that we present a further elaboration of Simon's HLT construct.

Simon (1995) offered the HLT as a way to explicate an important aspect of pedagogical thinking involved in teaching mathematics for understanding. In particular, it described how mathematics educators (i.e., teachers, researchers, and curriculum developers), oriented by a constructivist perspective and particular mathematics learning goals for students, can think about the design and use of mathematical tasks to promote mathematical conceptual learning. However, the description of the HLT stopped short of providing a framework for thinking about the learning process and the design or selection of mathematical tasks.

Recently, we articulated a mechanism—reflection on activity-effect relationships—for explaining mathematics concept development (Simon et al., in press). In the present article, we explain and exemplify how this mechanism can serve to elaborate the HLT construct by providing a framework for thinking about the hypothetical learning process and the design or selection of mathematical tasks. We begin by reviewing key aspects of the HLT construct and then review the mechanism of reflection on activity-effect relationships.

REVIEWING THE HYPOTHETICAL LEARNING
TRAJECTORY CONSTRUCT

An HLT consists of the goal for the students' learning, the mathematical tasks that will be used to promote student learning, and hypotheses about the process of the students' learning (Simon, 1995). Whereas the teacher's goal for student learning provides a direction for the other components, the selection of learning tasks and the hypotheses about the process of student learning are interdependent. The tasks are selected based on hypotheses about the learning process; the hypothesis of the learning process is based on the tasks involved. Underlying this construct are the following assumptions:

1. Generation of an HLT is based on understanding of the current knowledge of the students involved.
2. An HLT is a vehicle for planning learning of particular mathematical concepts.
3. Mathematical tasks provide tools for promoting learning of particular mathematical concepts and are, therefore, a key part of the instructional process.
4. Because of the hypothetical and inherently uncertain nature of this process, the teacher is regularly involved in modifying every aspect of the HLT.

REFLECTION ON ACTIVITY-EFFECT RELATIONSHIPS:
SUMMARIZING THE MECHANISM

Simon et al. (in press) articulated a mechanism for mathematics concept development grounded in a key aspect of constructivism—the construct of *assimilation*. According to Piaget, (1971, 1980), learners' assimilatory schemes structure their experiences and determine to what learners can attend. By introducing the notion of assimilation, Piaget asserted that learners cannot simply take in a new concept from the outside. Rather, learners construct new conceptions via assimilation into and accommodation of prior conceptions.

If one takes the notion of assimilation seriously, a problematic issue arises that is referred to as the "learning paradox" (Pascual-Leone, 1976). How can one explain learners' development of new conceptions without attributing to them prior conceptions that are as advanced as those to be learned (Bereiter, 1985)? Because learners are unable to structure their experience in ways that are not part of their current assimilatory conceptions, they are unable to perceive new conceptions in their experience of the world around them. For example, students, who have no concept of composite units of 10, will not see place-value relationships in base-10 blocks (Cobb, Yackel, & Wood, 1992)—relationships that are obvious to the teacher who selected the lesson involving the blocks. Therefore, if new conceptions cannot be taken in from the outside, how do learners develop more advanced conceptions based only on extant more primitive conceptions? The mechanism that we summarize here was postulated to addresses the learning paradox.

The mechanism, reflection on activity-effect relationship[1], is an elaboration of Piaget's reflective abstraction. Our articulation of the mechanism provides a description of how learners' goal-directed activity can lead to the generation of new (to the learners), more sophisticated conceptions. We now summarize the mechanism. The next section on using the mechanism provides an illustration of the mechanism within a particular conceptual domain.

The process begins with the learners setting goals. During instruction, the learners' goal setting is often in relationship to a task that has been posed by the teacher. The goals that the learners can set are a function of their current conceptions and related to the task at hand. The learners' goal should not be confused with the teacher's goal for student learning, a component of the HLT. For example, the teacher might introduce a game to foster a particular mathematical concept. The learners' goal would be to win the game, a goal that is distinctly different from the mathematics-learning goal identified by the teacher.

Having set a goal, the learners call on an available activity (or set of activities) in an effort to meet the goal. Working toward this goal, the learners attend to the effects of their goal-directed activity.

While attending to the effects of their activity (relative to their goal), the learners create mental records. The unit of experience that is recorded is an iteration of the activity linked to its effect. The learners sort and compare, whichleads to identifying patterns, that is, relationships among the activity and effects. This reflective abstraction of a new (to the learner) activity-effect relationship is the mechanism by which a new concept is constructed. Implied in this elaboration of reflective abstraction is that each of the components, creating records of experience, sorting and comparing records, and identifying patterns in those records, is an inborn mental ability and tendency (von Glasersfeld, 1995).

We add two points of clarification. First, following Piaget (1985), reflection, which is made up of these inborn mental abilities and tendencies, does not necessarily involve conscious thought. Thus, young children develop a concept of number, although little or no conscious thought is involved in the reflective process. Second, when learners are engaged in physical activity, the physical activity is associated with mental activity. It is the learners' mental activity, whether or not there is physical activity involved, that is the basis for conceptual learning.

---

[1]Description of the mechanism began in the context of Tzur's (1996, 1999, 2000) empirical studies on fraction learning.

To address the learning paradox, this explanation of conceptual learning must not attribute to the learner any conceptions that are beyond those available at the outset of the learning process. Let us reexamine the previously stated explanation from this perspective. First, following von Glasersfeld (1995), we assumed that learners have inborn abilities and tendencies (e.g., creating records of experience, sorting and comparing records, and identifying patterns in those records). Second, we assumed that, based on assimilatory conceptions available at the outset, learners have the ability to set a goal, select an activity that was learned previously, and monitor progress toward the goal. That learners are involved in goal-directed activity is central to addressing the learning paradox. Because the learners are attempting to achieve their goals, they use a particular activity and make adjustments to that activity for the purpose of achieving the goal. Thus, it is reasonable to claim that learners pay attention to their intentional variations in their goal-directed activity and to those effects of their activity that they can identify as a result of monitoring progress toward their goal. If we were to claim that learners can pay attention to all variations in their activity and to all effects of their activity that are perceivable by an observer, our explanation could be attacked on the basis of attributing to the learners conceptions beyond those available at the outset.

In the next section, we provide an example of lesson design to illustrate the mechanism and its role in generating HLTs.

USING THE MECHANISM TO GENERATE AN HLT

Successful mathematics lessons have always existed. The elaboration of the HLT construct discussed in this article provides a way to understand why successful lessons are successful and a framework for producing successful lessons and modifying unsuccessful lessons.

To illustrate the use of the mechanism of reflection on activity-effect relationships for generating an HLT, we explicate the development of a lesson involving equivalent fractions.[2] The example focuses only on the use of the mechanism for articulating a hypothetical learning process and selecting mathematical tasks. Not addressed are important aspects of instruction, such as mathematical discussions that take place in small and large groups. Before an HLT can be generated, the teacher must have an understanding of the students' mathematical conceptions. The following trajectory is based on an assumption that, at the outset, the students understand a fraction as a quantity (Simon, 2002; Tzur, 1999), know what the numerator and the denominator indicate, and given one of three representations—the fraction name, its symbol, and an area representation of the fraction—can generate

---

[2]A similar lesson sequence was discussed in Simon (2003). We anticipate that the reader will recognize the lesson as a reasonable attempt to meet the stated learning goals. Indeed, it is similar to what the reader would do. The purpose is to exemplify the underlying theoretical constructs.

the other two. It is assumed that they have a rudimentary understanding of equivalent fractions, specifically that they know that the same quantity can be represented by different fractions. It is also assumed that the students have an understanding of whole number multiplication and division and knowledge of multiplication/division number facts through 10×10.

The generation of an HLT begins with identifying a learning goal for students, which is based on knowledge of the students' current mathematical knowledge. The learning goal for this lesson[3] is for students to understand the quantitative relationships between a fraction and an equivalent fraction whose denominator is a multiple of the original fraction, that is, that the numerator and denominator of the equivalent fraction increase by the same factor. Although achievement of this goal will not afford a complete understanding of equivalent fractions, this limited goal is appropriate for a single HLT and sufficient to illustrate the role of the mechanism in task design.

We acknowledge that achievement of this goal for students' learning would not constitute a huge change in students' understanding. We use this example for three reasons. First, the example is likely to be accessible to a broad set of readers. More complex transitions would require specialized knowledge of the conceptual area involved. Second, a more significant conceptual transformation would require a set of HLTs all contingent on the effect of prior interventions—thus an example that would be much more complicated. Third, the goal for student learning in this example is often not realized during instruction on equivalent fractions (cf., Post, Cramer, Behr, Lesh, & Harel, 1993).

Once the learning goal has been identified, the creation of an HLT involves generating a hypothetical learning process in the context of a particular set of learning tasks. Where does one begin, given the interdependence of these two components? This is where the mechanism of reflection on activity-effect relationships comes into play. The emphasis on activity offers a starting point. The first question that we ask is "What activity, currently available to the students, might be the basis for the intended learning?" In some cases, this can be a difficult question. In this example, we identified the activity of subdividing fractional parts as the activity. (Students commonly subdivide fractional parts in their attempts to draw a particular number of parts, for example, drawing three parts and cutting them in half to make six parts.) The identification of the activity was part of a hypothesis that, on the basis of subdividing fractional parts, students could come to anticipate the logical necessity of the numerator and the denominator increasing by the same factor.

Because the mechanism of reflection on activity-effect relationships provides a framework for thinking about the learning process and the role of the task in that process, we have a framework for generating the HLT. We begin with a rough hy-

---

[3]We use "lesson" to designate a unit of instruction. It does not necessarily refer to one class period. A lesson may take place over multiple class periods.

pothesis for the learning process, guided by the mechanism of reflection on activity-effect relationships. It can be outlined as follows:

1. Using their current knowledge of fractions, the students will draw a diagram of the original fraction and subdivide the parts in the original fraction to create the number of parts in the new denominator.
2. Using their knowledge of whole number multiplication, the students will multiply the number of shaded parts (old numerator) by the number of subdivisions in each part to determine the new numerator.
3. Through reflection on their activity and its effects, the students will determine that the activity of multiplying the denominator (subdividing each part) causes the numerator (the number of shaded parts) to be increased by the same factor.

We next endeavor to design or select tasks that are likely to cause the students to set a goal, to call on the intended activity, and to reflectively abstract the intended concept. We generated the following tasks:

1. Draw a rectangle with 1/2 shaded. Draw lines on the rectangle so that it is divided into sixths. Determine how many sixths are in 1/2.
2. Draw a rectangle with 2/3 shaded. Draw lines on the rectangle so that it is divided into twelfths. Determine $2/3 = ?/12$.
3. Draw diagrams to determine the following:
   a. $3/4 = ?/8$
   b. $4/5 = ?/15$
   c. $3/4 = ?/20$

Having generated a set of tasks, we can now consider our hypothesis for the learning process in greater detail. We discuss the purpose of each task in light of our anticipation of the students' activity, the relevant effect of their activity, and reflection on the activity-effect relationship. The tasks were designed to encourage the setting of a goal of finding the new numerator and the specific activity of partitioning fractional parts. We anticipated that there were other ways that the students could use their diagrams (e.g., making a second diagram to show the second fraction) toward achieving their goal. Thus, we designed Tasks 1 and 2 to encourage students to transform the original diagram based on our anticipation that that activity would most reliably lead to the intended abstraction. Task 3 provided additional experience on which to reflect.

The purpose for creating the first three tasks was to promote the use of an activity sequence that could engender reflection on, and ultimately abstraction of, the intended activity-effect relationship. We intended the tasks to afford the possibility of the students abstracting this relationship from the activity without further intervention. Thus, there is the possibility for the students to spontaneously reflect on the activity-effect relationship—reflection that may not involve conscious thought. Using the activity-effect mechanism, we now detail the learning process that was conjectured for students' work with tasks 1 through 3.

Students' Activity Sequence

The students draw the original fraction, determine the number of subdivisions necessary for each fractional part, subdivide each part, and determine the number of subdivisions in the shaded area.

Effect of Students' Activity

The effect is the determination of the new denominator by multiplying the number of parts by the number of subdivisions per part. Our conjecture here is not that the students will determine a generalized relationship. Rather, they will spontaneously and with understanding perform the appropriate multiplication of the specific numbers involved in that diagram. For example, in finding the number of twentieths in 3/4, the students will notice that having subdivided each part into five subdivisions, they now have three groups of five subdivisions, that is, 3×5 twentieths in the shaded area (the new numerator).

We emphasize that from an observer's perspective, there is always the possibility of identifying numerous effects of the students' activity sequence. However, we are concerned with the students' perspective. We use two criteria for identifying an effect as being part of the learning process: (a) that the students could pay attention and would tend to pay attention to the effect, and (b) that the effect can contribute to the intended learning.

Reflection on Activity-Effect Relationship

Reflecting across their experiences of carrying out these tasks, we conjecture that the students will abstract a relationship between their activity and its effect, that subdividing the parts into x subdivisions results in x subdivisions in each shaded part or x times the original numerator (the new numerator) Thus, the students will form an anticipation that the number of shaded parts is increased by the same factor as the total number of parts. We emphasize here that the pattern that is recognized is in the activity-effect relation, not just the effects. This is not an observation of a numerical pattern resulting from a black-box process (for additional discussion of this point, see Simon, 2003). The students' engagement in the activity of subdividing and the resultant effect, the multiplication that determines the new numerator (an example of what Piaget, 2001, calls a "coordination of actions"), allows them to abstract the logical necessity of the relation.

Although we designed the first three tasks to promote reflection on a particular activity-effect relationship, we do not assume that all of the students in the class will reflectively abstract the relationship on the basis of these tasks. Therefore, we designed Task 4 to foster reflection on the activity-effect relationship and to make explicit the new mathematics that has been created, what is referred to in French Didactical Theory (Brousseau, 1997) as "situations of institutionalization."

4. Drawing diagrams to solve equivalent fractions problems is not much fun when the numbers get large. For the following do not draw a diagram. Rather describe what would happen at each step *if you were to draw a diagram*. Use that thinking to answer the following:
   a. 5/9 = ?/90
   b. 7/9 = ?/72

In Task 4, the students are asked to anticipate the effects of their original activity sequence. For some students this may cause them to engage in conscious reflection on the relationship of their activity and its effects in the context of the previous tasks. For those who have abstracted the relationship already, this problem can allow them to make that relationship explicit.

In analyzing our hypothesis of the students' knowledge following the first four tasks, we conclude that one additional aspect of the relationship between the equivalent fractions (one additional activity-effect relationship) is needed for the students to have an abstracted understanding of the type of fraction transformation involved in these problems. In the earlier tasks, for example 3/4 = ?/8, the students could see immediately that each of the four parts of the original fraction needs to be partitioned into two smaller parts. They may not think consciously about looking for a factor relating 4 and 8 or that division would be an appropriate operation for doing so. The mechanism of reflection on activity-effect relationship frames the generation of this second HLT.

For the purpose of fostering an abstraction about the division of the new denominator by the old denominator, we identify an activity that the students have used in the earlier problems, asking themselves how many subdivisions must be made in each part of the original fraction to make visible the new denominator. Task 5 is designed to engage students in that activity and to give them an opportunity for reflective abstraction. The students' goal is to find the new numerator under the conditions imposed by the tasks.

5. Without drawing a diagram, think in terms of cutting up a rectangle. Use a calculator to calculate the following. Write down each step that you do and the result you get. Justify each step in terms of how it is related to cutting up a rectangle.
   a. 16/49 = ?/147
   b. 13/36 = ?/324

Task 5 presents the students with pairs of denominators for which the factor difference between them is not obvious. Their first activity, looking for the number of subdivisions for each part, produces an effect: a division of the new denominator by the old denominator. Our conjecture is that the students can handle the problem "How many times do 49 parts need to be subdivided to make 147 subdivisions?" using their understanding of whole number division. What they have not developed (and we are attempting to foster) is an abstracted understanding of the multiplicative relationship between the new and old denominators. Thus, the students' activity is the mental operation of subdividing each of the original parts. The effect of the students' activity is the numerical operation of dividing the new denominator by the old denominator.

Over the course of tasks of this type, we conjecture that the students will abstract the activity-effect relationship, the multiplicative relationship between the new denominator and the original denominator. This, combined with their abstraction from the first four tasks (i.e., when they need to find the number of shaded parts in the new fraction, they multiply the number of subdivisions by the original numerator), completes their development of the abstracted relationship. In this way, they do not merely memorize a set of procedures. Rather they develop an understanding of the logical necessity of the relationships involved by building up a meaningful sequence of activities related with anticipated effects, that is, an image of subdivided figures coordinated with appropriate mathematical operations. Although the calculation method—divide the denominators and multiply the old numerator by their quotient—can become routine, the students maintain the ability to justify each step in terms of the activity-effect relationship on which it was built.

Finally, Task 6 has a parallel role to Task 4. It fosters additional reflection on the activity-effect relationship and asks the students to make their set of operations explicit.

    6.   Write a calculator protocol for calculating a problem of the form $a/b = ?/c$.

A couple of comments are in order regarding this sample lesson. As for any HLT, the lesson represents a conjecture—an informed starting place for creating appropriate lessons. There is no assurance of success and modifications would likely be needed based on interaction with the students. This example is not a first approximation for any particular group of students. The example is meant to focus on the sequence of tasks and how the sequence is structured on the basis of the mechanism reflection on activity-effect relationships. For students of a particular level, we might hypothesize the need for more experience with one or more of the tasks before they are ready to move on to the next. On the other hand, we do not assume that all students require experience with each of the six tasks. (See Simon et al., in press, for a more detailed description of the mechanism.)

DISCUSSION

Use of the Elaborated HLT

In this article, we have demonstrated how a mechanism that explains relationship between conceptual learning and mathematical tasks can provide a framework for the generation of HLTs. With this elaboration of the HLT, the selection of tasks is not left to intuition or trial and error. Rather, the mechanism offers a framework for thinking about how the task can promote the learning process. How might this elaboration of the HLT be used?

In curriculum design, this elaboration of the HLT could provide a framework for conceptualizing the creation of sets of lessons aimed at developing a new concept. Although recent curricula contain many effective lessons, such a framework could help structure the consistent generation of effective lessons. One program of research and development, Realistic Mathematics Education in the Netherlands (Gravemeijer, 1994), has benefited from building on an articulated framework. The framework we have described is consistent with the Realistic Mathematics Education framework and provides further elaboration of the relationship to students' prior conceptions and the mechanisms by which their conceptions are transformed.

In classroom teaching, this elaboration of the HLT can be useful, though it is not always indicated. Teachers who have been promoting inquiry mathematics (Richards, 1991) in their classrooms are aware that in some cases students, when given a problem or confronted with a cognitive conflict, can independently generate the necessary learning. In such cases, the mechanism we have presented could be used to analyze the students' learning processes. However, it might not be necessary for the teacher to engage in the complex planning that would result from using the framework. The most important use of the elaborated HLT would be for teaching concepts whose learning is problematic generally or for particular students. In such cases, greater understanding of learning processes and how they can be supported, of which the work here is but one step, is essential for developing a theoretical basis for dealing with difficult pedagogical problems. Our elaboration of the HLT can provide a framework not only for the design of lessons, but also for the modification of lessons that do not achieve their goal. In such cases, each component of the framework can be reexamined in light of observations made of the students.

The elaborated HLT can also be useful for research using a teaching experiment methodology (Steffe & Thompson, 2000). This methodology has been used to make key contributions to the knowledge base on learners' development of particular mathematical concepts. Researchers' success in using the methodology has been dependent on the researchers' competence in creating effective learning tasks. This elaboration of the HLT can provide an explicit framework for both the generation and modification of teaching experiment tasks.

Challenges in the Use of the Elaborated HLT

Perhaps the greatest challenge in using the elaborated HLT for lesson design is the identification of a mental activity that can lead to the intended concept. Taking a mathematical concept and reconceptualizing it as an activity-effect relationship can be difficult. Whereas, there is not necessarily only one activity that can be used as a basis for constructing a new concept, considerable pedagogical progress can be made when a useful activity is identified. An example is the research program of Steffe and his colleagues (cf., Olive, 1999; Steffe, 2002; Tzur, 1999) in their teaching experiments on fraction concept development. The program was based on the hypothesis that learners' iteration schemes, developed in the context of whole numbers, could serve as the basis for their construction of fraction concepts. However, it is important to note that a more micro-level analysis of their work is also appropriate, an analysis of the activity-effect relationships related to the development of particular fraction schemes. Whereas the particular activity sequences generally involve iteration, it is the differences among these sequences that contribute to an account for how each scheme is built on the foundation of prior schemes.

A second challenge derives from using this explanatory mechanism to account for successful learning in the context of particular mathematical tasks. Here the challenge is to identify from among the myriad of activities and effects that can be identified by the observer, a particular coupling of activity and effect that can account for the learning. For us, a necessary, although perhaps not sufficient, aspect of this work is to think as the learner—to try to find a reasonable explanation that is consistent with the assimilatory schemes of the learner at each point in the process.

Ongoing Work

For the more difficult to learn mathematical concepts, additional empirical work is needed to understand better how these concepts can be developed. Historically, research studies that report conceptions of students at particular points have been more common. However, studies of learning of concepts (transitions from one level of understanding to another) are necessary and lacking in many conceptual areas. The elaboration of the HLT can assist in structuring both the planning and analysis aspects of the research.

We are currently working on an additional theoretical aspect of conceptual learning. Consistent with the work of others (e.g., Dubinsky, 1991; Piaget, 1954; Pirie & Kieren, 1994; Sfard, 1991), our empirical work has led to distinctions among stages of conceptual learning. We are in the process of describing how the mechanism of reflection on activity-effect relationships can be applied differentially at the different stages to account for the transitions between stages. This work leads to differentiation, stage-specificity of HLTs and of the tasks used in generating HLTs. A preliminary discussion of this work appeared in Tzur and Simon (1999).

ACKNOWLEDGMENTS

This research was supported by the National Science Foundation (REC 9600023) and by a fellowship from The National Academy of Education—The Spencer Foundation. The opinions expressed do not necessarily reflect the views of the foundation.

The authors acknowledge the involvement of Karen Heinz and Margaret Kinzel in the work reported on in this article. We thank Karen Heinz and Koeno Gravemeijer for their comments on an earlier draft of the article.

REFERENCES

Ainley, J., & Pratt, D. (2002). Purpose and utility in pedagogical task design. In A. Cockburn & E. Nardi (Eds.), *Proceedings of the Twenty Sixth Annual Conference of the International Group for the Psychology of Mathematics Education: Vol. 2* (pp. 17–24). Norwich, UK: School of Education and Professional Development, University of East Anglia.

Bell, A. (1993). Principles for the design of teaching. *Educational Studies in Mathematics, 24,* 5–34.

Bereiter, C. (1985). Toward a solution to the learning paradox. *Review of Educational Research, 55,* 201–226.

Brousseau, G. (1997). *Theory of didactical situations in mathematics.* Dordrecht, The Netherlands: Kluwer.

Cobb, P., Yackel, E., & Wood, T. (1992). A constructivist alternative to the representational view of mind in mathematics education. *Journal for Research in Mathematics Education, 23,* 2–33.

Dubinsky, E. (1991). Reflective abstraction in advanced mathematical thinking. In D. Tall (Ed.), *Advanced mathematical thinking.* Dordrecht, The Netherlands: Kluwer.

Goldin, G., & McClintock, E. (Eds.) (1985). *Task variables in mathematical problem solving.* Philadelphia: Franklin Institute Press.

Gravemeijer, K. (1994). *Developing realistic mathematics education.* Culemborg, The Netherlands: Tecgbuoress.

Krainer, K. (1993). Powerful tasks: A contribution to a high level of acting and reflecting in mathematics instruction. *Educational Studies in Mathematics, 24,* 65–93.

Lappan, G., & Briars, D. (1995). How should mathematics be taught? In I. M. & Carl (Eds.), *Seventy-five years of progress: Prospects for school mathematics* (pp. 115–156). Reston, VA: National Council of Teachers of Mathematics.

National Council of Teachers of Mathematics. (2000). *Principles and standards for school mathematics.* Reston, VA: Author.

Olive, J. (1999). From fractions to rational numbers of arithmetic: A reorganization hypothesis. *Mathematical Thinking and Learning, 1,* 279–314.

Pascual-Leone, J. (1976). A view of cognition from a formalist's perspective. In K. F. R. J. A. Meacham (Ed.), *The developing individual in a changing world: Vol. I. Historical and cultural issues.* The Hague, The Netherlands: Mouton.

Piaget, J. (1954). *The construction of reality in the child* (M. Cook, Trans.). New York: Ballantine.

Piaget, J. (1971). *Biology and knowledge*. Chicago: University of Chicago Press.

Piaget, J. (1980). *Adaptation and intelligence*. Chicago: University of Chicago Press.

Piaget, J. (1985). *The equilibration of cognitive structures: The central problem of intellectual development*. Chicago: University of Chicago Press.

Piaget, J. (2001). *Studies in reflecting abstraction*. Sussex, England: Psychology Press.

Pirie, S., & Kieren, T. (1994). Growth in mathematical understanding: How can we characterise it and how can we represent it? *Educational Studies in Mathematics, 26,* 165–190.

Post, T., Cramer, K., Behr, M., Lesh, R., & Harel, G. (1993). Curriculum implications of research on the learning, teaching, and assessing of rational number concepts. In T. Carpenter & E. Fennema (Eds.), *Research on the learning, teaching, and assessing of rational number concepts*. Hillsdale, NJ: Lawrence Erlbaum and Associates, Inc.

Richards, J. (1991). Mathematical discussions. In E. von Glasersfeld (Ed.), *Radical constructivism in mathematics education* (pp. 13–51). Dordrecht, The Netherlands: Kluwer.

Sfard, A. (1991). On the dual nature of mathematical conceptions: Reflections on processes and objects as different sides of the same coin. *Educational Studies in Mathematics, 22,* 1–36.

Simon, M. (1995). Reconstructing mathematics pedagogy from a constructivist perspective. *Journal for Research in Mathematics Education, 26,* 114–145.

Simon, M. (2002). Focusing on critical understandings in mathematics. In D. Mewborn, P. Sztajn, D. White, H. Weigel, R. Bryant, & K. Nooney (Eds.), *Proceedings of the Twenty-fourth Annual Meeting of the North American Chapter of the International Group for the Psychology of Mathematics Education: Vol. 2* (pp. 991–998). Athens: University of Georgia.

Simon, M. (2003). Logico-mathematical activity versus empirical activity: Examining a pedagogical distinction. In N. Pateman, B. J. Dougherty, & J. Zilliox (Eds.), *Proceedings of the 27th Conference of the International Group for the Psychology of Mathematics Education: Vol. 4* (pp. 183–190). Honolulu, HI: XXX.

Simon, M., Tzur, R., Heinz, K., Smith, M., & Kinzel, M. (in press). Explicating a mechanism for conceptual learning: Elaborating the construct of reflective abstraction. *Journal for Research in Mathematics Education.*

Smith, M. S., & Stein, M. K. (1998). Selecting and creating mathematical tasks: From research to practice. *Mathematics Teaching in the Middle School, 3,* 344–50.

Steffe, L. P. (2002). A new hypothesis concerning children's fractional knowledge. *Journal of Mathematical Behavior, 102,* 1–41.

Steffe, L., & Thompson, P. (2000). Teaching experiment methodology: Underlying principles and essential elements. In A. K. R. Lesh (Ed.), *Handbook of research design in mathematics and science education* (pp. 267–306). Hillsdale, NJ: Lawrence Erlbaum Associates, Inc.

Tzur, R. (1999). An integrated study of children's construction of improper fractions and the teacher's role in promoting that learning. *Journal for Research in Mathematics Education, 30,* 390–416.

Tzur, R. (2000). An integrated research on children's construction of meaningful, symbolic, partitioning-related conceptions, and the teacher's role in fostering that learning. *Journal of Mathematical Behavior, 18*(2), 123–147.

Tzur, R., & Simon, M. (1999). Postulating relations between levels of knowing and types of tasks in mathematics teaching: A constructivist perspective. In F. Hitt & M. Santos (Eds.), *Twentieth-First Annual Meeting North American Chapter of the International Group for the Psychology of Mathematics Education: Vol. 2* (pp. 805–810). Cuernavaca, Mexico: ERIC.

van Boxtel, C., van der Linden, J., & Kanselaar, G. (2000). Collaborative learning tasks and the elaboration of conceptual knowledge. *Learning and Instruction, 10,* 311–330.

van Glasersfeld, E. (1995). *Radical constructivism: A way of knowing and learning*. Washington, DC: Falmer.

MATHEMATICAL THINKING AND LEARNING, 6(2), 105–128

Local Instruction Theories as Means of Support for Teachers in Reform Mathematics Education

Koeno Gravemeijer

Freudenthal Institute & Department of Educational Research
Utrecht University

This article focuses on a form of instructional design that is deemed fitting for reform mathematics education. Reform mathematics education requires instruction that helps students in developing their current ways of reasoning into more sophisticated ways of mathematical reasoning. This implies that there has to be ample room for teachers to adjust their instruction to the students' thinking. But, the point of departure is that if justice is to be done to the input of the students and their ideas built on, a well-founded plan is needed. Design research on an instructional sequence on addition and subtraction up to 100 is taken as an instance to elucidate how the theory for realistic mathematics education (RME) can be used to develop a local instruction theory that can function as such a plan. Instead of offering an instructional sequence that "works," the objective of design research is to offer teachers an empirically grounded theory on how a certain set of instructional activities can work. The example of addition and subtraction up to 100 is used to clarify how a local instruction theory informs teachers about learning goals, instructional activities, student thinking and learning, and the role of tools and imagery.

INTRODUCTION

In the 1960s and 1970s theories for instructional design were in vogue in the edu-cational-research community. The most well-known design theories from that pe-riod are probably Gagné's *Principles of Instructional Design* (Gagné & Briggs, 1974). Since then, the interest for instructional design has faded away. More re-

---

Requests for reprints should be sent to Koeno Gravemeijer, Freudenthal Institute & Department of Educational Research, Utrecht University, Tiberdreef 4 3561 GG Utrecht, The Netherlands. E-mail: k.gravemeijer@fi.uu.nl

cently, however, a renewed interest can be noticed, especially in communities of mathematics educators. This relates to the current reform efforts in mathematics education. Constructivism formed one of the catalysts of this reform movement. Various interpretations of constructivism fueled the belief that mathematics education should capitalize on the inventions by the students. This in part can explain why there initially was little interest in instructional design within this reform movement. We can even put it in stronger terms: For many, instructional design was seen as incompatible with mathematics education that put the students' own ideas and input at the forefront. This gradually changed, and the insight is growing that mathematics education that aims to capitalize on the input of the students requires thorough planning. However, this would mean a different kind of planning than envisioned in traditional instructional design strategies.

The instructional design principles of the 1960s and 1970s do not fit reform mathematics instruction. The main problem is that the older design principles take as their point of departure the sophisticated knowledge and strategies of experts to construe learning hierarchies. Following a *task analysis* approach, the performance of the expert is taken apart and laid out in small steps, and a learning hierarchy is constituted that describes what steps are prerequisite and in what order these steps should be acquired. The result is a series of learning objectives that can make sense from the perspective of the expert, but not necessarily from the perspective of the learner. Further, there is little room for personal input from the learner.

What is needed for reform mathematics education is a form of instructional design supporting instruction that helps students to develop their current ways of reasoning into more sophisticated ways of mathematical reasoning. For the instructional designer this implies a change in perspective from decomposing ready-made expert knowledge as the starting point for design to imagining students elaborating, refining, and adjusting their current ways of knowing.

This change of perspective encompasses both a change in pedagogy and a change in curriculum; reform mathematics asks for a specific classroom culture and discourse but it also asks for another curriculum and corresponding instructional activities. In the current discussion on reform in mathematics education, the former often get the most attention. The reform pedagogy is elaborated in terms of classroom culture, social norms, mathematical discourse, mathematical community, and a stress on inquiry and problematizing. Without denying the importance of this aspect of reform, it could be necessary to draw the attention to the curriculum counterpart of this innovative pedagogy.

The central problem in reform in mathematics teaching is the well-known tension between the openness toward the students' own constructions and the obligation to work toward certain given endpoints. Or as Deborah Ball (1993) noted:

> How do I create experiences for my students that connect with what they know and care about but also transcend the present? How do I value their interests and also con-

nect them to ideas and traditions growing out of centuries of mathematical exploration and invention? (p. 375)

It is a question of how the teacher could proactively support students' mathematical development, as Cobb (1996) noted. The pedagogy mentioned ensures openness toward students' own constructions. Working toward given end goals asks for more. It asks for what Simon (1995) called "hypothetical learning trajectories." The teacher has to envision how the thinking and learning in which the students could engage as they participate in certain instructional activities relates to the chosen learning goal. Simon emphasized the hypothetical character of these learning trajectories; the teachers must analyze the reactions of the students in light of the stipulated learning trajectory to find out to what extant the actual learning trajectory corresponds with what was envisioned. Based on this information the teacher has to construe new or adapted instructional activities in connection with a revised learning trajectory.

The example Simon (1995) worked out shows that designing hypothetical learning trajectories for reform mathematics is no easy task. We can, therefore, ask ourselves what kind of support can be given to teachers. It is clear that we cannot rely on fixed, ready-made, instructional sequences, because the teacher will continuously have to adapt to the actual thinking and learning of his or her students. Thus, it seems more adequate to offer the teacher some framework of reference, and a set of exemplary instructional activities that can be used as a source of inspiration.

This is exactly what is aimed for in the Dutch tradition of developing realistic mathematics education (RME). Here the objective is to design support materials by trying to construe learning paths along which students could reinvent conventional mathematics. Such a learning path is paved with instructional activities that can function as stepping stones in this conjectured reinvention process. The conceptualization of these learning paths is of the same character as that of Simon's (1995) learning trajectories. Significant differences with hypothetical learning trajectories, however, are the duration of the learning process, and the "situatedness" in a specific classroom, or more to the point, the lack thereof. To emphasize the distinction, I reserve the term *hypothetical learning trajectories* for the planning of instructional activities in a given classroom on a day-to-day basis, and I use the term *local instruction theories* to refer to the description of, and rationale for, the envisioned learning route as it relates to a set of instructional activities for a specific topic (e.g., addition and subtraction up to 20, area, fractions, etc.).

The relation between hypothetical learning trajectories and local instruction theories can be elucidated with Simon's (1995) travel metaphor. In terms of a travel metaphor, the local instruction theory offers a "travel plan," which the teacher has to transpose into an actual "journey" with his or her students. The idea is that the teachers use their insight in the local instruction theory to choose instructional activities and to design hypothetical learning trajectories for their own students. In my view, local instruction theories can never free the teachers from having to design hypothetical learning trajectories for their own classroom. Nevertheless, I would argue, that using a local instruction theory as a framework of reference could enhance the quality of the learning trajectories.

This, in a sense, is the main point of this article: Externally developed local instruction theories are indispensable for reform mathematics education. It is unfair to expect teachers to invent hypothetical learning trajectories without any means of support. In addition, it can be argued that without them, the chances to reconcile openness toward students' own contributions and aiming for given end goals are very slim.

To develop local instruction theories to support teachers, a theory is needed on how to help students' construct mathematical ideas and procedures. The point of departure here is that RME offers such a theory and that design research is the appropriate method for developing local instruction theories. I elucidate this in the following way. I start with a description of design research as a method for developing local instruction theories. Next I use this as a background to elucidate the RME theoretical framework on the basis of an exemplary local instruction theory. I complement this by highlighting the very aspects in which the local instruction theory goes beyond the level of an instructional sequence in terms of a series of instructional activities.

DESIGN RESEARCH

Local instruction theories are developed in what is called developmental research (Gravemeijer, 1994, 1998) or design research (Gravemeijer & Cobb, 2001). The core of this type of research is formed by classroom teaching experiments that center on the development of instructional sequences and the local instructional theories that underpin them. In the course of a classroom teaching experiment, the research team develops sequences of instructional activities that embody conjectures about the course of students' learning. To this end, the designer conducts an anticipatory thought experiment by envisioning both how proposed instructional activities can be realized in interaction in the classroom and what mental activities students can engage in as they participate in them. Analyses of the actual process of students' mental activities when they participate in the instructional activities as constituted in the classroom can then provide valuable information that can be used to guide the revision of the instructional activities. The rationale for the instructional sequence can be conceived as a local instructional theory that underpins a prototypical instructional sequence (Gravemeijer, 1994, 1998).

The design research at the Freudenthal Institute grew out of the desire to develop mathematics education that corresponds with Freudenthal's (1973, 1991) ideal of "mathematics as an human activity." According to Freudenthal, students should be given the opportunity to reinvent mathematics by mathematizing—mathematizing subject matter from reality and mathematizing mathematical subject matter. In both cases, the subject matter that is to be mathematized should be experientially real for the students. That is why this approach is named RME. One of the core principles of RME is that mathematics can and should be learned on one's own authority and through one's own mental activities. That is to say, students should experience the process by which new mathematics is learned as a reinvention process in which they themselves play an active role. Moreover, they should develop sufficient intellectual autonomy (Kamii, Lewis, & Livingstone Jones, 1993) to only accept new mathematical knowledge of which they can judge the validity themselves. Within the RME research community, the question is asked what mathematics education would have to look like to fulfill the previously discussed educational philosophy by experimenting with mathematics education in practice and by reflecting on this experimental practice. This reflection leads to the development of an educational theory, and this theory feeds back into new experiments. This implies that the resulting theory that I call "a domain-specific instruction theory for realistic mathematics education" (or RME theory, for short) is always under construction.

In conjunction with the theory, a research method emerged in the Netherlands that was labeled developmental research.[1] Similar approaches emerged elsewhere, for instance, under the names of design experiments (e.g. Brown, 1992; Cobb, McClain, & Gravemeijer, 2003) and design research (e.g. Edelson, 2002).[2] In this article, I use the latter term, design research, which seems to be more common than the Dutch label of developmental research.

Preliminary Design

Design research that focuses on the development of local instruction theories basically encompasses three phases: developing a preliminary design, conducting a teaching experiment, and carrying out a retrospective analysis. The first phase starts with the clarification of mathematical learning goals, combined with anticipatory thought experiments in which one envisions how the teaching-learning process can be realized in the classroom. This first step results in the explicit formulation of a *conjectured local instruction theory* that is made up of three components: (a) learning goals for students, (b) planned instructional activities and the tools that

---

[1] Accidentally, the same label, developmental research, is used for another Dutch research approach. Instructional design is also at the heart of this research approach; however, the goal is not to develop domain specific instruction theories but to develop improved design theories (van den Akker, 1999).

[2] It should be noted that the description of developmental research presented here also builds on what is learned in the collaboration of the author with Paul Cobb and Kay McClain at Vanderbilt University (see also Gravemeijer & Cobb, 2001).

will be used, and (c) a conjectured learning process in which one anticipates how students' thinking and understanding could evolve when the instructional activities are used in the classroom.

This conjectured local instruction theory is open to adaptations on the basis of input of the students and assessments of their actual understandings. The theory also reflects the importance of anticipating the possible process of their learning as it could occur when planned instructional activities are used in the classroom. The manner in which a conjectured local instruction theory is construed can be described as "theory-guided bricolage" (Gravemeijer, 1994) because it resembles the manner of working of an experienced "tinkerer," or "bricoleur." The design researcher follows a similar approach using and adapting existing ideas and materials, but the way in which selections and adaptations are made is guided by a theory—in our case, RME theory. RME theory offers three design heuristics, denoted as *guided reinvention*, *didactical phenomenology*, and *emergent modeling* (which I discuss in more detail later). These design heuristics help the research team in designing a possible learning route together with a set of potentially useful instructional activities that fit this learning route. More specifically, this implies that the researchers think through what mental activities of the students can be expected when they engage in the instructional activities and how those mental activities can help the students to develop the envisioned mathematical insights. In the teaching experiment, those conjectures are put to the test.

The Teaching Experiment

This process of anticipating and testing is, in fact, an iterative process that resembles Simon's (1995) "mathematics teaching cycle." The actual enactment of the instructional activities in the classroom enables the researchers to investigate whether the mental activities of the students correspond with the ones they anticipated. The insights gained in this manner and the experience with the instructional activities as such form the basis for the design or modification of subsequent instructional activities and for new conjectures about what mental activities of the students can be expected. In this manner, instructional activities are tried, revised, and designed on a daily basis during the teaching experiment. This cyclic process of thought experiments and instruction experiments (Freudenthal, 1991) forms the backbone of the design research method employed in the teaching experiment (see Fig. 1).

Even though the researchers carry out thought experiments and instruction experiments on a daily basis, the goal of the research team is not to prepare the next day's instructional activity, but to develop a well-considered and empirically grounded local instruction theory. The term local instruction theory is coined to convey the intention of offering more than a description of a learning route, or the corresponding instructional activities. In addition to these two, a local instruction theory also includes a rationale. In contrast with traditional instructional design re-

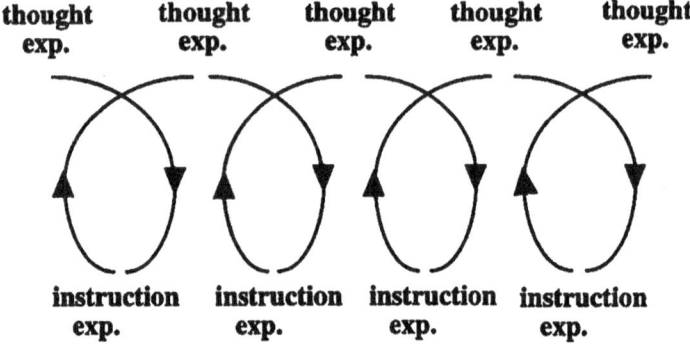

thought exp.   **thought exp.**   **thought exp.**   **thought exp.**   **thought exp.**

instruction exp.   **instruction exp.**   **instruction exp.**   **instruction exp.**

FIGURE 1 Design research, a cumulative cyclic process.

search, the objective of design research is not to offer an instructional sequence that "works," but to offer the user an empirically grounded theory on how the researchers think that a certain set of instructional activities could work (cf. NCTM Research Advisory Committee, 1996).

This is in line with the earlier described notion of a travel plan; it does not seem plausible that an instructional sequence could be enacted in the exact same manner in a variety of classrooms—especially if we expect the teacher to adapt to the students' thinking. In contrast, individual teachers can use a local instructional theory as a framework of reference for the design of hypothetical learning trajectories that fit the actual needs of their students.

In the design research project, the mathematical teaching cycles serve the development of the local instruction theory. In fact, there is a reflexive relation between the thought and instruction experiments and the local instruction theory that is being developed. On one hand, the conjectured local instruction theory guides the thought and instruction experiments, and on the other hand, the micro instruction experiments shape the (conjectured) local instruction theory (Fig. 2).

To be able to adjust the envisioned instructional activities on a daily basis, it is desirable that the researchers be present in the classroom every day while the teaching experiment is in progress. The ongoing analyses of individual children's activity and of classroom social processes inform new anticipatory thought experiments in the course of which conjectures about possible learning trajectories are frequently revised. As a consequence, there is often an almost daily modification of local learning goals and instructional activities.

This focus on the ongoing process of experimentation emphasizes that ideas and conjectures are modified while interpreting students' reasoning and learning in the classroom. The empirical data on the activities of the students are interpreted in light of the theoretical framework of RME. In addition to this, the interpretive

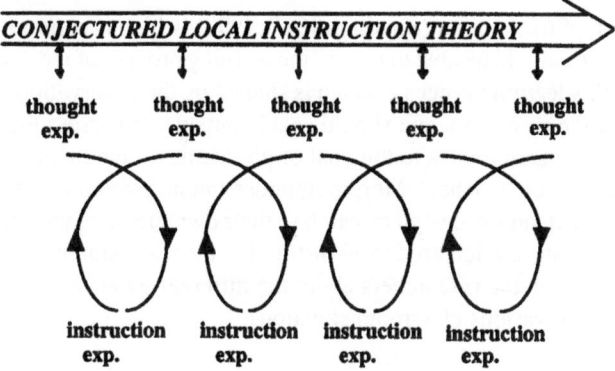

FIGURE 2 Reflexive relation between theory and experiments.

framework of Cobb and Yackel (1996) can be used to help the researchers make sense of classroom events.

Retrospective Analysis

The results of design experiments cannot be linked to pretest and posttest results in the same direct manner as is common in standard formative evaluation, because the proposed local instruction theory and prototypical instructional sequence will differ from what is tried in the classroom. Because of the cumulative interaction between the design of the instructional activities and the assembled empirical data, the intertwinement between the two has to be unraveled to pull out the optimal instructional sequence in the end. For it does not make sense to include activities that did not match their expectations, but the fact that these activities were in the sequence will have affected the students. Therefore, adaptations will have to be made when the nonfunctional, or less functional, activities are left out.

Consequently, the instructional sequence is put together as a reconstruction of a set of instructional activities, which are thought to constitute the effective elements of the sequence. This reconstruction of the optimal sequence is based on the deliberations and the observations of the research team. In this manner, the result of a developmental research experiment is a well-considered and empirically grounded rationale for the envisioned learning route in connection with the proposed set of instructional activities.[3] Methodologically, this result has to be justified by the learning process of the research team. In relation to this, we can refer to a method-

[3]The retrospective analysis can spark ideas that surpass what is tried out in the classroom. This can create the need for a new developmental research project, starting with a new conjectured local instruction theory. In this manner, subsequent teaching experiments can become part of a series of macro cycles of experimentation and revision.

ological norm of trackability (Smaling, 1987, 1992), which is common in ethnography; outsiders should be able to retrace the learning process of the research team. Insight into this learning process, which is shaped by the observations and the deliberations of the research team, should enable outsiders to assess the viability of the results. This again relates to the goal of producing viable theories that can be adapted by classroom teachers. Moreover, understanding the *how* and *why* enables the teachers to extend the design research to their own practice, within which they experiment with the conjectured local instruction theory. Actually, feedback from teachers can inform the researchers about the different ways in which the theory can be adapted to various classroom situations.

EXEMPLARY LOCAL INSTRUCTION THEORY

In the following, I discuss the core elements of a local instruction theory on the basis of an instructional sequence that is developed in a teaching experiment in Nashville, Tennessee by Cobb, Gravemeijer, McClain, and Stephan of Vanderbilt University (Stephan, Bowers, Cobb, & Gravemeijer, 2000). But, before focusing on the design of the instructional sequence and the corresponding local instruction theory, I want to stress the importance of the classroom culture that is essential for the enactment of such an instructional sequence. To realize a problem-centered, or inquiry-based, learning process, certain classroom social norms (Cobb & Yackel, 1996) need to be established. Such social norms can include expectations and obligations regarding explaining and justifying solutions, attempting to make sense of explanations given by others, indicating agreement and disagreement, and questioning alternatives in situations in which a conflict in interpretations or solutions has become apparent.

In addition to this, certain socio-math norms must be established to create the opportunity for the students to evaluate mathematical progress.

The Design

The goal of the instructional sequence I use as an example is to foster the use of flexible mental computation strategies for addition and subtraction up to 100.[4] In designing a conjectured local instruction theory, we can build on the experience gathered in several decades of developmental/design research at the Freudenthal Institute and elsewhere. This research effort has resulted in a domain-specific instruction theory that is grounded in numerous concrete elaborations of the RME approach (Gravemeijer, 1994; Streefland, 1990; Treffers, 1987). By interpreting

---

[4]Actually, there was a dual goal: linear measurement and flexible arithmetic (e.g., Stephan, Bowers, Cobb, & Gravemeijer, 2004). In this article, however, I limit myself to the arithmetic part.

this domain-specific instruction theory as an instructional design theory, we can point to three design heuristics mentioned previously: guided reinvention (Freudenthal, 1973), didactical phenomenological analysis (Freudenthal, 1983), and emergent modeling (Gravemeijer, 1999).

The *design principle of guided reinvention* is the key principle of RME. According to Freudenthal (1973), the students should be given the opportunity to experience a process similar to the process by which a given piece of mathematics was invented. For the designer, this implies that a route has to be mapped out that allows the students to invent the intended mathematics by themselves. To do so, the researcher starts with imagining a route by which he or she could have personally arrived at this outcome. In doing so, the designer can take both the history of mathematics and students' informal solution procedures as sources of inspiration.

According to the reinvention principle, the goal of the local instruction theory on addition and subtraction up to 100 is not to teach the students solution strategies in the form of ready-made techniques. Instead, the goal is to help the students develop similar solution methods on their own accounts. A plausible model, then, is to assume that students initially base their computations on their familiarity with certain number relations. When a problem such as $29 + 5 = \ldots$ has to be solved, number relations involving 29 and 5 can come to mind. A student could think of $9 + 1 = 10, 29 + 1 = 30, 5 + 4 = 9, 4 + 1 = 5, 5 + 5 = 10$, and so forth and try to use these to solve the problem at hand. One option would be to combine $29 + 1 = 10$ and $4 + 1 = 5$ to conclude that you can take 1 of the 5, add that 1 to the 29 to get 30, and add the remaining 4 to the 30 to get 34. Then, from an observer's point of view, it could look like the student is using a building-up-to-10 strategy. For the student, however, this strategy might not be on the horizon yet. Only after reflecting on substantial experience with similar problems, the student could start to notice a pattern and construe the building-up-to-10s strategy. Even then, it could take a while before the student starts to use this strategy as an a priori guidance for choosing a solution procedure.

Thus, the choice of guided reinvention as our point of departure is intertwined with the way we frame our goals. The instructional goal is not to teach the students a set of strategies. Instead, our primary goal is for the students to develop a framework of number relations that offers the building blocks for flexible mental computation.

Having said that, we still can ask ourselves, what solution procedures—or what use of number relations—to aim for. Research on the solution procedures students use to solve addition and subtraction problems up to 100 shows that those procedures fall in two broad categories (Beishuizen, 1993), which we may denote as *splitting* and *counting*.

A task like $44 + 37$, for instance, can be solved in the following manner,

by splitting 10s and 1s:
$44 + 37 = \ldots$; $40 + 30 = 70$; $4 + 7 = 11$; $70 + 11 = 81$; or

by counting in jumps:

$44 + 37 = \ldots; 44 + 30 = 74; 74 + 7 = 81;$ or

$44 + 37 = \ldots; 44 + 6 = 50; 50 + 10 = 60; 60 + 10 = 70; 70 + 10 = 80; 80 + 1 = 81,$ or

via some other combination of jumps of 10s and 1s.

Beishuizen (1993) found that procedures based on splitting 10s and 1s leads to more errors than solution procedures that are based on counting on and counting back. Moreover, the latter type leaves room for a wide variety of solution procedures and offers more opportunities for curtailing and inventing shortcuts. Counting-by-jumps therefore fits best the type of instructional sequence we aim for.

It can further be noted that, as the example shows, decuples are used as reference points in this counting-by-jumps strategy. In relation to this, decuples also play a central role in framework of number relations that we want the students to develop.

The RME-guided reinvention heuristic is connected with mathematizing; the students invent by mathematizing. The idea is that the students not only mathematize contextual problems—to make them accessible for a mathematical approach—but also mathematize their own mathematical activity, which brings their mathematical activity at a higher level. Freudenthal (1971) characterized mathematizing as a form of organizing, which is also a key element of his didactical phenomenology (Freudenthal, 1983) that constitutes the second design heuristic.

Didactical phenomenology is grounded in a phenomenology of mathematics, within which the focus is on the relation between a mathematical "thought thing" (*nooumenon*) and the "phenomenon" it describes and analyses, or, in short, organizes.

Phenomenology of a mathematical concept, a mathematical structure, or a mathematical idea means, in my terminology, describing the *nooumenon* in relation to the *phainomena* of which it is the means of organizing, indicating which phenomena it is created to organize and to which it can be extended, how it acts upon these phenomena as a means of organizing, and with what power over these phenomena it endows us. (Freudenthal, 1983, p. 28)

In a didactical phenomenology, this relation of mathematical thought thing (concept, structure, or idea) and phenomenon is analyzed from a didactical point of view; the focus is on how the relation is acquired in a learning-teaching process. Freudenthal (1983) contrasted his approach with the (then) conventional approach of trying to concretize abstract concepts (in an "embodiment"). In the latter approach, he concluded, one puts the cart before the horse by teaching abstractions by concretizing them.

What a didactical phenomenology can do is to prepare the converse approach: starting from those phenomena that beg to be organized and, from that starting point, teaching the learner to manipulate these means of organizing. Didactical phenomenology is to be called in to develop plans to realize such an approach. In the didactical phenomenology of length, number, and so on, the phenomena organized by length, number, and so on are displayed as broadly as possible. (Freudenthal, 1983, p. 23)

The didactical phenomenological analysis can orient the researchers toward applied problems that can be suitable as points of impact for a process of progressive mathematization. So, rather than looking around for material that concretizes a given concept, the didactical phenomenology suggests looking for phenomena that might create opportunities for the learner to constitute the mental object that is being mathematized by that very concept.

In relation to a phenomenology of numbers, Freudenthal (1983) noted, "numbers organize the phenomenon of quantity," whereas "the phenomenon 'number' is organized by means of the decimal system" (p. 28). He worked this out in more detail for addition, starting with the lowest level, which is to combine two sets—as in 5 cars and 3 cars or 5 marbles and 3 marbles. However, he argued, problems arise when the addition is not plainly recognizable as the union of two sets, as is the following case: John has 5 marbles, and Pete has 3 more. How many does Pete have? Instead of uniting two given sets, the students must consider the imaginary set of Pete as split into two sets, and reason from there.

Next to those situations, in which addition is not plainly recognizable as the union of two sets, there are also situations in which it is less natural to speak of sets consisting of 5 and 3 elements, such as 5 steps (of stairs) and 3 steps, 5 days and 3 days, or 5 kilometer and 3 kilometer. With those spatial or temporal phenomena one cannot speak of a union of two unstructured sets. Instead, counting is used to organize magnitudes, in which measuring the magnitude is articulated by the natural multiples of a unit. Continuous phenomena are made discrete by a one-to-one mapping of the successive intervals on a sequence of points that follow each other in space or time, in a process that in turn suggests a counting process. In line with this sequential character, the results of additions of magnitudes are obtained by counting on. In relation to this, Freudenthal (1983) pointed to the close relation between cardinal and ordinal numbers: "5 + 3 is defined cardinally, but from olden times it has been calculated ordinarily" (p. 99). The result of 5 + 3 is obtained by starting with the mental 5, and counting on, 6, 7, 8.

From this phenomenological analysis, Freudenthal (1983) concluded, "Counting can and must immediately be transferred from discrete quantities, represented by sets, to magnitudes" (p. 101). He recommended the number line as a device that visualizes magnitudes and, at the same time, the natural numbers. The number line also lends itself to express *more* or *less* as directions. In this manner, the number line, or two parallel number lines, can also be used to visualize the problem of Pete who has three marbles more than John.[5] This reference to the use of the number line brings us to the issue of models and modeling, for example, the next design heuristic.

The third heuristic, the *emergent-modeling design heuristic*, assigns a role to models that differs from the role of ready-made models as embodiments of abstract concepts mentioned earlier. Instead of trying to concretize abstract mathematical knowledge, the objective is to try to help students model their own informal mathematical activity. The aim is that the model with which the students model their own informal mathematical activity gradually develops into a model for more formal mathematical reasoning. However, the model I am referring to is more an overarching concept than one specific model. In practice, the model in the emergent modeling heuristic is actually shaped as a series of consecutive symbolizations or tools[6] that can be described as a cascade of inscriptions or a chain of signification. From a more global perspective, these tools can be seen as various manifestations of the same model. So when I speak of a shift in the role of the model in the following, I am talking about the model on a more general level. On a more detailed level, this transition can encompass various tools that gradually take on different roles.

The label *emergent* refers both to the character of the process by which models emerge within RME and to the process by which these models support the emergence of formal mathematical ways of knowing. According to the emergent-models design heuristic, the model first comes to the fore as a model of the students' situated informal strategies. Then, over time the model gradually takes on a life of its own. The model becomes an entity in its own right and starts to serve as a model for more formal, yet personally meaningful, mathematical reasoning. In relation to this, we discern four different types or levels of activity (Gravemeijer, 1999, 2002):

1. activity in the task setting,
2. referential activity,
3. general activity, and
4. more formal mathematical reasoning.

---

[5]Although the design of the instructional sequence under discussion is in line with the previously discussed elaboration of a didactical phenomenology of number, it should be noted that Freudenthal's elaboration was not the actual source for the design. The research team came to similar conclusions on the basis of didactical phenomenological considerations in connection with earlier design experiments. The research team's didactical-phenomenological deliberations build on the observation that students tend to come up with a wide variety of counting solutions when confronted with linear-type context problems (e.g., Vuurmans, 1991). In addition, a closer look at counting strategies shows us that these strategies rely on integrating the cardinal aspect of number (quantity) and the ordinal aspect of number (position/rank). Most addition and subtraction problems concern quantities, whereas the solution procedures consist of moving up and down the number sequence. We argue that it is important that the students connect the first and the latter. This then inspired us to try to integrate measurement and the empty number line.

[6]I use the word "tools" as a generic term in the following discussion, encompassing also symbolizations, or inscriptions.

These levels of activity underline that the model is grounded in students' under-standings of paradigmatic, experientially real task settings. In other words, the model emerges as situation-specific imagery. This implies that initially, at the ref-erential level, the model is meaningful for the students because it signifies for them the activity in the task setting to which it refers. General activity begins to emerge as the students start to reason about the mathematical relations involved. As a con-sequence, the model loses its dependency on situation-specific imagery and gradu-ally develops into a model that derives its meaning from a framework of mathemat-ical relations that is being construed in the process. The transition from model-of to model-for coincides with a progression from informal to more formal mathe-matical reasoning that involves the creation of new mathematical reality, which is thought of as consisting of mathematical objects (Sfard, 1991) within a framework of mathematical relations. The level of more formal activity is reached when the students no longer need the support of models.

As an aside, it can be noted that this transition cannot be pinned down to one specific symbolization or tool. Instead, there is a gradual change in the way the stu-dents perceive and use tools as their personal framework grows.

Several authors have proposed the use of the number line as a means of support for addition and subtraction up to 100 (Freudenthal, 1983; Treffers & de Moor, 1990; Whitney, 1988). For me, the objective to help students make a connection between the cardinal and the ordinal aspect forms the main argument to introduce the number line as a tool. From an expert's point of view, a number line integrates both the cardinal aspect (line segment) and the ordinal aspect (point). In addition to this, the number line offers a way of symbolizing that fits nicely the various count-ing strategies—by describing the subsequent counting steps as arcs on an empty number line (Gravemeijer, 1994, 1999). I speak of an empty number line because this number line is empty except for the numbers that are actually needed. The stu-dents add these to the number line as a part of the solution process (see Fig. 3).

The interpretation of the number line, however, is not self-evident. For the stu-dents, it does not speak for itself what the marks on the number line signify. The hash marks might signify either cardinal or ordinal numbers and not necessarily a synthe-sis of the two. Exactly for that reason, Whitney (1988) and Treffers (Treffers & de Moor, 1990) let their introduction of the number line be preceded by activities with a

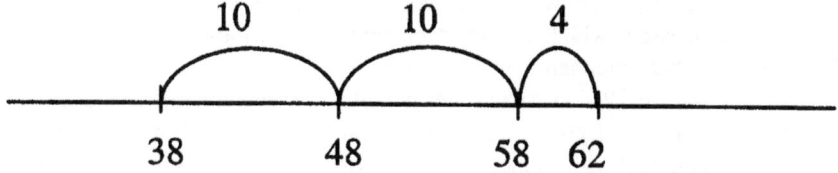

FIGURE 3 Solving 38 + 24 on the empty number line.

bead string. This bead string consists of 100 beads, colored in groups of 10. While executing various counting tasks, students find out that the decimal structure can be used to solve tasks like, "Count 38 and add 24 more. Which number do you get?" This solution procedure is being modeled with arcs on the empty number line. The indispensability of this kind of imagery proved itself in a design experiment where the bead string was skipped (Cobb, Gravemeijer, Yackel, McClain, & Whitenack, 1997). The students did not have the means to resolve among themselves whether a hash mark with a number, say 38 for example, should be thought of as signifying 38 objects (e.g. candies or beads) or the thirty-eighth object.

I concur with Freudenthal's (1983) arguments to ground addition and subtraction in linear measurement, as measuring presents itself as a natural alternative for counting on a bead string. Conceptual understanding of measurement requires that students interpret the activity of measuring as the accumulation of distance (Thompson & Thompson, 1996). Similarly, a number on a ruler would have to signify the total measure of the distance measured. Speculating on the genesis of the ruler in history, one can take the view that the ruler came about as a curtailment of iterating a measurement unit. So, the ruler can be thought of as a model of iterating some measurement unit, whereas the empty number line can function as a model for more sophisticated mathematical reasoning in the context of mental computation strategies with numbers up to 100. The connection between the two can be made by building on the relation between iterating measurement units as accumulating distances and a cardinal interpretation of positions on the number line. This is truly a model-of/model-for transition if it coincides with a shift in the student's view of numbers as referents of distances to numbers as mathematical entities. This shift involves a transition from viewing numbers as tied to identifiable objects or units (i.e., numbers as constituents of magnitudes, such as 38 feet) to viewing numbers as mathematical objects (e.g., 38). For the student, a number viewed as a mathematical object still has quantitative meaning, but this meaning is no longer dependent upon its connection with identifiable distances or with specified countable items. Instead, numbers viewed as mathematical objects derive their meaning from their place in a network of number relations.

The enacted instructional sequence.   With the help of the previously discussed elaboration of the design heuristics, we developed a preliminary design of the instructional sequence, which was worked out in the teaching experiment in Nashville, Tennessee, which is well-documented in various publications (e.g., Gravemeijer, 1999; Stephan, 1998; Stephan et al., in press; Stephan, Cobb, Gravemeijer, & Estes, 2001). Space does not allow for an elaborate account of all findings here. Instead, I give a brief description of the enacted instructional sequence, supplemented with elements of the retrospective analysis that offer essential background theory for teachers. With the latter, I want to highlight the importance of offering teachers more than a set of instructional activities. Note that I present a somewhat smoothed description of the enacted sequence in which encountered detours are left out.

The sequence starts with a story of a country where the king's foot serves as the measurement unit and the king does all the measuring. The students follow the king's example by pacing various items in the classroom (heel-to-toe). In the sequel to this story, an alternative is sought to the king doing all the measuring. After some discussion a footstrip of five feet is introduced and the students start measuring with this footstrip. Later, a new scenario is introduced, which is about Smurfs (small blue dwarfs) who measure with food cans. Those food cans happen to have the same size as the Unifix cubes that are available in the classroom. The students measure objects by stacking Unifix cubes until they reach the required length.[7] As the story unfolds, measuring with individual cubes is exchanged for measuring with a "Smurfbar," consisting of 10 Unifix cubes, which implies coordinating 10s and 1s. In the context of the story, this Smurfbar is invented to free the Smurfs from the task of carrying numerous food cans with them. To get rid of the obligation to carry food cans completely, a paper "10-strip" is made to replace the Smurfbar. Finally, ten 10-strips are pasted together to construe a measurement strip of 100 cubes long. Next, a significant step is taken, when the activity of measuring with the measurement strip is followed by tasks about incrementing, decrementing, and comparing lengths. Here the students have to take a length as a given and use the measurement strip as a means of support for solving the problems. Gradually, the counting strategies that the students use to solve those problems are replaced by forms of arithmetical reasoning that build on the measurement strip. Instead of counting the difference between 34 and 56, for instance, the students may reason $34 + 6 = 40$, $40 + 10 = 50$, and $50 + 6 = 56$, so the difference is $6 + 10 + 6 = 22$. When the students have reached this stage, the empty number line is introduced as a means of scaffolding and communicating the arithmetical solution procedures that the students use. As a last step, the empty number line notation is generalized to support arithmetical reasoning in contexts other than measuring.

The local instruction theory.   The previously discussed design experiment resulted in the local instruction theory that is summarized in Table 1, which gives an overview of the potential tool use, the corresponding imagery and mathematical activity, and topics of mathematical discourse. I do not lay out this instruction theory in detail, nor do I underpin it with an account of the retrospective analysis of the learning process of the classroom community. All this is described in detail in Stephan et al. (in press). Instead, I use elements of the retrospective analysis as a basis for an elaboration of two more fundamental dimensions of the researchers' theory on how the instructional activities can work, which teachers must come to grips with to be able to design hypothetical learning trajectories for their own

---

[7]Because Unifix cubes can be clicked together, it is rather easy to make a solid stack of cubes.

TABLE 1

Overview of the Proposed Role of Tools in the Instructional Sequence

Tool	Imagery	Activity/T-a-s Interests	Potential Mathematical Discourse Topics
Feet (heel to toe)		Measuring	
Masking tape	Record of activity of pacing	Reasoning about activity of pacing	Focus on covering distance
Footstrip	Record of pacing (builds on masking tape) (Form/function shift: using a record of pacing as a tool for measuring)	Measuring with a "big step" of five = measuring by iterating a collection of paces	Measuring as divorced from activity of measuring; Structuring distance in collections of 5s and 1s
Smurf cans	Stack of Unifix cubes signifies result of iterating	Measuring by creating a stack of Unifix cubes	Builds on measuring divorced from activity of iterating
Smurf bar	Signifies result of iterating	Measuring by iterating a collection of 10 Unifix cubes; Structuring distance into measures of 10s and 1s	Accumulation of distances; Coordinating measuring with 10s with measuring by 1s
10-strip	Signifies measuring 10s and 1s with the Smurf bar	Measuring by iterating the 10-strip, and using the strip as a ruler for the 1s	Accumulation of distances; Coordinating 10s & 1s
Measurement strip	Signifies measuring with 10 strip/ Starts to signify result of measuring (Form/function shift: inscription developed for measuring is used for scaffolding and communicating)	(1) Measuring: strip alongside item; counting by 10s and 1s => reading of endpoint (2) Reasoning about spatial extensions (results of measuring have become entities in and of themselves)	Distance seen as already partitioned; extension already has a measure; Part-whole reasoning/quantifying the gaps between two or more lengths; Shift in focus: focus on number relations; developing and using emergent framework of number relations
Empty number line	Signifies reasoning with measurement strip	Means of scaffolding & means of communicating about reasoning about number relations	Numbers as mathematical entities (numbers derive their meaning from a framework of number relations); Various arithmetical strategies

Note. Reprinted with permission from Stephan, M., Bowers, J., Cobb, P., & Gravemeijer, K. (Eds.), *Supporting students' development of measuring conceptions: Analyzing students learning in social context. Journal for Research in Mathematics Education Monograph No. 12.* Copyright 2003 by the National Council of Teachers of Mathematics. All rights reserved.

classrooms. The first concerns an empirically grounded theory on how the students are expected to make sense of acting with new tools and how this relates to preceding activities. The second concerns an empirically grounded theory on the students' conceptual development in relation to the relevant mathematical concepts.

Tools and Imagery

In RME design, the emergent model and the corresponding series of tools function as the backbone of the intended reinvention process. Ideally, the students should invent the necessary tools for themselves. This, however, is not really feasible. We take care, however, that the students are involved in the invention process. This can be done by a careful introduction of each new tool according to the following set up: Each new tool has to come to the fore as a solution to a problem (e.g., how to measure more efficiently). First, the students are given the opportunity to think about and discuss possible solutions to that problem, then the students are asked to evaluate whether the new tool offers an acceptable solution to the problem. In this manner, the students experience an involvement in the invention process even though they do not invent the tools for themselves. In this manner, we try to ensure that the tools emerge in a sense from the activity of the students. In addition, we make sure that the use of new tools is grounded in some imagery for the students. That is to say, there has to be some history in the learning process of the student that renders meaning to the activity with a new tool. I already pointed to this issue of imagery in relation to the interpretation of the hash marks on the number line. The brief sketch of the enacted instructional sequence reveals the following series of tools that the students use: feet, footstrip, Smurf cans, Smurf bar, 10-strip, measurement strip, and empty number line.

The first link concerns pacing with one's feet and measuring with the footstrip. In practice, the connection between the two was mediated by the construction of a record of the activity of pacing. The teacher made this record to facilitate the discussion of the different ways the students were counting the placement of their feet, when measuring by pacing heel-to-toe. The teacher placed pieces of masking tape at the beginning and end of each pace, which enabled her—and the students—to point to the various paces after the fact. With help of this record, the students discussed whether the first foot should be counted. Thanks to this history, this record signified the activity of pacing for the students. The footstrip that is introduced later builds on this notion of a record of pacing as the footstrip, too, can be seen as a record of pacing. But then a form/function shift (Saxe, 1991) takes place when the students start using this record of pacing as a tool for measuring. However, as measuring with the footstrip builds on the imagery of the activity of pacing with individual feet, putting down the footstrip can signify pacing heel-to-toe for the students.

When the students start measuring with the Unifix cubes in the Smurf scenario, there is no direct link to earlier activities in terms of imagery. The learning history of the students, however, does play a role because the experience of pacing and measuring with the footstrip enables the students to consider a stack of Unifix cubes as a result of measuring, an observation that was underscored by the students' inability to do this when we tried to introduce the Unifix cubes too quickly.

The transition from measuring with individual cubes to measuring with the Smurf bar is similar to the transition between pacing and measuring with the footstrip. Next, measuring with the 10-strip builds on the history of measuring with the Smurf bar. Thanks to this history, measuring with the 10-strip signifies measuring 10s and 1s with the Smurf bar for the students. In turn, measuring with the measurement strip builds on the imagery of measuring with the 10-strip. Actually, the students initially count by 10s and 1s on the measurement strip to establish the length of a item. Gradually, however, a position on the measurement strip starts to signify the result of measuring. Then, we meet another form/function shift, when the tasks shift from measuring to reasoning about measures, and the measurement strip is used as a means for scaffolding and communicating ways of reasoning. Finally, drawing arcs on the empty number line is introduced as an alternative means for scaffolding and communicating ways of reasoning with the measurement strip.

The main idea behind the design of this cascade of tools is that the way in which the students act and reason with each tool builds on their activity with earlier ones. This build up is to ensure that the students have a meaningful way of acting with the tools, because they can rely on the imagery of acting with earlier, already familiar tools. From this perspective, it is essential that the teachers who want to reenact the sequence come to grips with the researchers' empirically grounded theory on how reasoning with one tool builds upon the other.

Potential Mathematical Discourse Topics

A significant feature of the instruction, described previously, is that the agenda of the designers differs from the immediate goals of the tasks. The immediate goal for the students is, for instance, to figure out how long something is or what the difference between two lengths is. But the instructional objective is to create a situation that gives rise to various solution strategies, which in turn lend themselves to a discussion on significant mathematical issues. It is therefore important that teachers, who intend to reenact this sequence, understand these potential mathematical discourse topics and their relation to the intended mathematical development of the students. The potential mathematical discourse topics listed in Table 1 are taken from the doctoral thesis of Michelle Stephan (1998), which discerned a series of mathematical practices that reflect the way in which taken-as-shared ways of reasoning, arguing, and tool use evolved as the sequence was enacted. Those mathematical practices both encompass the way of acting and reasoning with tools and the conceptual understanding involved.

The instructional sequence starts with students measuring objects by pacing. However, for some of the students, the goal appeared to be just to count the number of steps it took them to reach the end of the item. This was inferred from the fact that some students did not count the placement of the first foot (when the heel was aligned with the beginning of the item measured). They started counting, "one," with the placement of the second foot, whereas others started their counting with the first foot. What we are aiming for is that the students come to see measuring as covering amounts of space. To reach this goal, the teacher has to make the two different ways of measuring a topic of discussion (whether you have to count the first foot). In such a discussion, the students can start to realize that it is not just a matter of convention; instead, if one does not count the first foot, an amount of the item would not be measured. In this manner, students can come to see the goal of measuring as covering amounts of space—as was the case in the experimental classroom.

With the activity of measuring with a footstrip of five paces, many students ran into problems when the space that was being covered by the footstrip extended past the physical extension of the measured item. For them, apparently, measuring was tied to the physical act of placing a footstrip, and they could not mentally cut the footstrip when needed. The instructional goal here is that the space to be measured takes priority over the measurement activity and becomes independent of activity for the students. Whole-class discussions on concrete instances are needed to create opportunities for students to articulate how the extended footstrip can be mentally cut.

The activity of measuring with a Smurf bar in the Smurf scenario showed that, for some students, the curtailment of counting by individual cubes was based only on a number word relation. For instance, when the second iteration of the Smurf bar would extend beyond the item measured, they would count the cubes past the first iteration as "21, 22, 23 and so forth", instead of, "11, 12, 23,...." For them, "20" seemed to be the number word associated with the second placement of the Smurf bar rather than the amount of space covered by 20 cubes. Although what is aimed for is that the students realize that as "20" signifies the length covered by 20 cubes, 21, 22, and 23 must extend beyond the length whose measure was 20. In other words, the students have to come to grips with coordinating measuring with 10s with measuring with 1s. To achieve this, the latter has to become a topic of discussion in the classroom.

In subsequent activities, the Smurf bar is replaced by paper 10-strips and next ten 10-strips are taped together to make a measurement strip of 100 cans long. Here, the students may initially measure with the measurement strip by laying the strip down alongside the item and counting by 10s and 1s until they reach the endpoint of the item. Gradually, however, the students curtail their activity of counting up on the measurement strip and find that the length of an item can be measured by laying down the measurement strip alongside the item and simply reading off the numeral corresponding to the position of the farthest endpoint. To do this insightfully, they have to conceive an extension as already having a measure, independent of the activity of measuring.

The next set of activities involves tasks such as comparing the lengths of two items and figuring out the difference with the help of the measurement strip. This is the first instructional activity in which the students do not measure an item that is physically present. The mathematically significant issue here is the quantification of a gap between two numbers. In the teaching experiment, some students counted the spaces between the two numbers whereas others counted the lines. To overcome this problem, the teacher made the different ways of quantifying the gaps a topic of discussion and asked the students to explain what each line or space signified to them. As a result of such discussions, the method of counting spaces to specify the measure of the spatial extension between two lengths became taken-as-shared.

A next step that is aimed for is that the students gradually replace the method of literally counting spaces by arithmetical reasoning. Again the teacher plays an important role by stimulating discussions on the different ways of establishing the number of spaces.

Finally, the empty number line is introduced as a means of describing and scaffolding various forms of arithmetical reasoning. When making the transition from the measurement strip to the number line, it is essential that the students differentiate between the activity of measuring and the activity of representing arithmetical strategies. On the empty number line the goal is for students to express how they would, for instance, increment 64 with 28. For example, by first measuring 64, then adding six 1s, which would get one to 70, then measuring two times 10, which would result in 80 and 90, respectively, and finally adding two 1s, which adds up to 92. When describing this method, it would be sufficient to show that when starting at 64, add 6, arrive at 70; then add 10, arriving at 80; another 10, arriving at 90; and 2, arriving at 92. To strive for an exact proportional representation of all the jumps would severely hamper a flexible use of the number line. Therefore we must make sure that the students are aware of the distinction between the ruler as a measurement tool and the empty number line as a means of describing solution procedures. Thus when they make drawings, the intention of the students should not be to make a schematic drawing of a measuring device, but to make a drawing that shows their arithmetical reasoning.

What is expected is, that in the course of the sequence, a shift is taking place in which the student's view of numbers transitions from referents of distances to numbers as mathematical entities. As argued before, this shift involves a transition from viewing numbers as tied to identifiable objects or units to viewing numbers as entities on their own that derive their meaning from a framework of number relations. This framework of number relations, then, offers the basis for flexible mental computation strategies for addition and subtraction up to 100, which was our instructional goal.

CONCLUSION

The main issue of this article is what instructional design has to offer to reform mathematics education, whereas classical instructional design theories do not fit mathematics education that tries to capitalize on the inventions of the students. The classical approach of task analysis results in a breakdown of the mathematical content in a hierarchy of small learning objectives that have to be mastered in a fixed sequence. This sequence is to be followed independent of the input or interest of the students; the only variation is one in speed and reteaching. A final drawback of the analytically defined learning objectives is that the students cannot see the relevance until they have reached the end of the process.

Still, I argue, if justice is to be given to the input of the students and their ideas built on, a well-founded plan is needed. In this respect, I point to the proactive role of the teacher in establishing an appropriate classroom culture, in choosing and introducing instructional tasks, organizing group work, framing topics for discussion, and orchestrating discussion. Following Simon (1995), this implies designing, enacting, assessing, and revising hypothetical learning trajectories in an iterative series of mathematical teaching cycles.

I use the example of the local instruction theory on addition and subtraction to show that design research can help teachers by developing viable local instruction theories, which can be used by classroom teachers to construe hypothetical learning trajectories that fit the characteristics and actual situations of their own classrooms. I highlight the word *theory* because, in contrast with traditional design theories, the emphasis is not on an elaborated instructional sequence with detailed directions for the teacher, but on the theory that underpins a possible instructional sequence—a theory of which we claim offers an empirically grounded theory on how the instructional sequence can work. The examples of the theory behind the way the various tools build on each other and the theory on how the conceptual development of the students can be supported by exploiting potential mathematical topics for discussion shed light on the theoretical framework that teachers need to make informed decisions in the classroom. In line with the RME theory that inspired the design, this enables teachers to design instruction that helps students to develop their current ways of reasoning into more sophisticated ways of mathematical reasoning.

REFERENCES

Ball, D. (1993). With an eye on the mathematical horizon: Dilemmas of teaching elementary school mathematics. *Elementary School Journal, 93,* 373–397.

Beishuizen, M. (1993). Mental strategies and materials or models for addition and subtraction up to 100 in Dutch second grades. *Journal for Research in Mathematics Education, 24,* 294–323.

Brown, A. L. (1992). Design experiments: Theoretical and methodological challenges in creating complex interventions in classroom settings. *Journal of the Learning Sciences, 2,* 63–112.

Cobb, P. (1996). Theories of mathematical learning and constructivism: A personal view. In G. Kadunz, H. Kautschitsch, G. Ossimitz, & E. Schneider (Eds.), *Shriftenreihe didaktikder mathematik (Vol. 23): Trends und perspectiven* (pp. 61–84). Vienna: Verlag Holder-Pichler-Tempsky.

Cobb, P., Gravemeijer, K., Yackel, E., McClain, K., & Whitenack, J. (1997). Mathematizing and symbolizing: The emergence of chains of signification in one first-grade classroom. In D. Kirschner & J. A. Whitson (Eds.), *Situated cognition theory: Social, semiotic, and neurological perspectives* (pp. 151–233). Hillsdale, NJ: Lawrence Erlbaum Associates, Inc.

Cobb, P., McClain, K., & Gravemeijer, K. (2003). Learning about statistical covariation. *Cognition and Instruction, 21,* 1–78.

Cobb, P., & Yackel, E. (1996). Constructivist, emergent, and sociocultural perspectives in the context of developmental research. *Educational Psychologist, 31,* 175–190.

Edelson, D. C. (2002). Design research: What we learn when we engage in design. *Journal of the Learning Sciences, 11,* 105–121.

Freudenthal, H. (1971). Geometry between the devil and the deep sea. *Educational Studies in Mathematics, 3,* 413–435.

Freudenthal, H. (1973). *Mathematics as an educational task.* Dordrecht, The Netherlands: Reidel.

Freudenthal, H. (1983). *Didactical phenomenology of mathematical structures.* Dordrecht, The Netherlands: Reidel.

Freudenthal, H. (1991). *Revisiting mathematics education.* Dordrecht, The Netherlands: Kluwer Academic.

Gagné, R. M., & Briggs, L. J. (1974). *Principles of instructional design.* New York: Holt, Rinehart & Winston.

Gravemeijer, K. (1994). *Developing realistic mathematics education.* Utrecht, The Netherlands: CD-β Press.

Gravemeijer, K. (1998). Developmental research as a research method. In J. Kilpatrick & A. Sierpinska (Eds.), *Mathematics education as a research domain: A search for identity (An ICMI study)* (Vol. 2, pp. 277–295). Dordrecht, The Netherlands: Kluwer Academic.

Gravemeijer, K. (1999). How emergent models may foster the constitution of formal mathematics. *Mathematical Thinking and Learning, 1,* 155–177.

Gravemeijer, K. (2002). Preamble: From models to modeling. In K. Gravemeijer, R. Lehrer, B. Van Oers, & L. Verschaffel (Eds.), *Symbolizing, modeling and tool use in mathematics education* (7–12). Dordrecht, The Netherlands: Kluwer Academic.

Gravemeijer, K. P. E., & Cobb, P. (2001, April). *Designing classroom-learning environments that support mathematical learning.* Paper presented at the Conference of the American Educational Research Association, Seattle, WA.

Kamii, C., Lewis, B. A., & Livingstone Jones, S. (1993). Primary arithmetic: Children inventing their own procedures. *Arithmetic Teacher, 41*(4), 200–203.

NCTM Research Advisory Committee. (1996). Justification and reform. *Journal for Research in Mathematics Education, 27,* 516–520.

Saxe, G. (1991). *Culture and cognitive development: Studies in mathematical understanding.* Hillsdale, NJ: Lawrence Erlbaum Associates, Inc.

Simon, M. A. (1995). Reconstructing mathematics pedagogy from a constructivist perspective. *Journal for Research in Mathematics Education, 26,* 114–145.

Smaling, A. (1987). *Methodologische objectiviteit en kwalitatief onderzoek* [Methodological objectivity and qualitiative research]. Lisse, The Netherlands: Swets en Zeitlinger.

Smaling, A. (1992). Varieties of methodological intersubjectivity—The relations with qualitative and quantitative research, and with objectivity. *Quality & Quantity, 26,* 169–180.

Sfard, A. (1991). On the dual nature of mathematical conceptions: Reflections on processes and objects as different sides of the same coin. *Educational Studies in Mathematics, 22,* 1–36.

Stephan, M., Bowers, J., Cobb, P., & Gravemeijer, K. (Eds.) (2004). Supporting students' development of measuring conceptions: Analyzing students' learning in social context. *Journal for Research of Mathematics Education Monograph, No. 12.*

Stephan, M., Cobb, P., Gravemeijer, K., & Estes, B. (2001). The role of tools in supporting students' development of measuring conceptions. In A. Cuoco (Ed.), *The roles of representation in school mathematics* (pp. 63–76). Reston, VA: National Council of Teachers of Mathematics.

Stephan, M. L. (1998). *Supporting the development of one first-grade classroom's conceptions of measurement: Analyzing students' learning in social context.* Unpublished doctoral dissertation, Vanderbilt University, Nashville, TN.

Streefland, L. (1990). *Fractions in realistic mathematics education, a paradigm of developmental research.* Dordrecht, The Netherlands: Kluwer Academic.

Thompson, A., & Thompson, P. (1996). Talking about rates conceptually. Part II: Mathematical knowledge for teaching. *Journal for Research in Mathematics Education, 27,* 2-24.

Treffers, A. (1987). *Three dimensions. A model of goal and theory description in mathematics education: The Wiskobas Project.* Dordrecht, The Netherlands: Reidel.

Treffers, A., & de Moor, E. (1990). *Proeve van een nationaal programma voor het reken-wiskundeonderwijs op de basisschool. Deel II: Basisvaardigheden en cijferen* [A specimen of a national program for mathematics education in primary school. Part II. Basic skills and written algorithms]. Tilburg, The Netherlands: Zwijsen.

van den Akker, J. (1999). Principles and methods of development research. In J. van den Akker, R. M. Branch, K. Gustafson, N. Nieveen & T. Plomp (Eds.), *Design approaches and tools in education and training* (pp. 1–14). Boston: Kluwer Academic.

Vuurmans, A. C. (1991). *Rekenen tot honderd* [Additions and subtraction up to hundred]. Hertogenbosch, The Netherlands: KPC.

Whitney, H. (1988). *Mathematical reasoning, early grades.* Unpublished manuscript, Princeton University.

MATHEMATICAL THINKING AND LEARNING, 6(2), 129–162

On the Construction of Learning Trajectories of Children: The Case of Commensurate Fractions

Leslie P. Steffe

Department of Mathematics Ed
University of Georgia

Learning trajectories are presented of 2 fifth-grade children, Jason and Laura, who participated in the teaching experiment, Children's Construction of the Rational Numbers of Arithmetic. 5 teaching episodes were held with the 2 children, October 15 and November 1, 8, 15, and 22. During the fourth grade, the 2 children demonstrated distinctly different partitioning schemes—the equi-partitioning scheme (Jason) and the simultaneous partitioning scheme (Laura). At the outset of the children's fifth grade, it was hypothesized that the differences in the 2 schemes would be manifest in the children's production of fractions commensurate with a given fraction. During the October 15 teaching episode, Jason independently produced how much 3/4 of 1/4 of a stick was of the whole stick as a novelty, and it was inferred that he engaged in recursive partitioning operations. An analogous inference could not be made for Laura. The primary difference in the 2 children during the teaching episodes was Laura's dependency on Jason's independent explanations or actions to engage in the actions that were needed for her to be successful in explaining why a fraction such as 1/3 was commensurate to, say, 4/12.

Simon (1995b) introduced the concept of a hypothetical learning trajectory to "refer to the teacher's prediction as to the path by which learning might proceed. It is hypothetical because the actual learning trajectory is not knowable in advance. It characterizes an expected tendency" (p. 135). He elaborated, "It is meant to underscore the importance of having a goal and rationale for teaching decisions and the hypothetical nature of such thinking" (p. 136). What I am concerned with in this article is the sense in which a learning trajectory is considered as hypothetical and

---

Requests for reprints should be sent to Leslie P. Steffe, University of Georgia, Department of Mathematics Ed, 105 Aderhold Hall, Athens, GA 30602. E-mail: LSteffe@coe.uga.edu

how one can be constructed. In a reaction to Simon's paper, Beatriz D'Ambrosio and I (Steffe & D'Ambrosio, 1995) emphasized designing a learning space that is based, at least in part, on a working knowledge of students' mathematics in the construction of a learning trajectory. In a reciprocal reaction, Simon (1995a) amplified our emphasis in his comments:

> They have (appropriately, in my opinion) emphasized the teacher's construction of models of the students' mathematics as one of the most important foci of models of teaching based on constructivism and agreed that the teacher's knowledge is constantly being constructed as she interacts with students as they construct knowledge. (p. 162)

But, who is "the teacher," and whose responsibility is it to construct learning trajectories? In radical constructivism, the principle of self-reflexivity[1] compels me to consider my own knowledge as constantly being constructed as I interact with students as they construct knowledge. So, rather than consider my own knowledge of how children learn mathematics as "good enough," and thereby consider the construction of learning trajectories as the responsibility of practicing teachers, I consider the construction of learning trajectories in the context of "the idea of worlds being constructed, or even autonomously invented, by inquirers who are simultaneously participants in those same worlds" (Steier, 1995, p. 71). Through the construction of learning trajectories that are coproduced by children, it is possible to construct *learning trajectories of children* that include an account of one's own ways and means of acting and operating as a teacher.

ELEMENTS IN THE CONSTRUCTION OF LEARNING TRAJECTORIES OF CHILDREN

The construction of learning trajectories of children is one of the most daunting but urgent problems facing mathematics education today. It is also one of the most exciting problems because it is here that we can construct an understanding of children's mathematics and how we as teachers can profitably affect that mathematics. By building an understanding of children's mathematical concepts and operations and how a teacher can engage children to bring forth changes in those concepts and operations, a vision of children's mathematics education can emerge in which children engage in productive mathematical learning and teachers engage in productive mathematical teaching. When learning to engage in productive mathematical teaching, I believe that we must immerse ourselves deeply and intensely in the experience of bringing forth productive mathematical learning on the part of students in teaching experiments (Steffe & Thompson, 2000).

---

[1]Self-reflexivity involves applying one's epistemological tenets first and foremost to oneself.

A learning trajectory of children includes a model of their initial concepts and operations, an account of the observable changes in those concepts and operations as a result of the children's interactive mathematical activity in the situations of learning, and an account of the mathematical interactions that were involved in the changes. Such a learning trajectory of children is constructed during and after the experience in intensively interacting with children. Prior to engaging children in interactive mathematical activity, I engage in thought experiments of the sort that produces what Simon (1995b) referred to as a hypothetical learning trajectory. As part of a three-year teaching experiment concerned with children's construction of fractional knowledge (Steffe & Olive, 1990), we taught two children, Jason and Laura, together during their fifth grade in school with the intent of investigating how they can construct commensurate fractions. I use the phrase *commensurate fractions* rather than *equivalent fractions* to refer to a particular kind of equivalences that children make on transforming, say, 1/3 into 5/15 without necessarily being able to produce a class of such fractions.

Prior to entering the fifth grade, the two children had constructed different partitioning schemes that proved to be essential in their construction of commensurate fractions during their fifth grade in school. I turn now to explaining these partitioning schemes because I regard them as the children's initial concepts and operations with respect to the construction of commensurate fractions.

THE CHILDREN'S INITIAL PARTITIONING SCHEMES

An Account of Jason's Equi-Partitioning Scheme

Jason had constructed what I refer to as an equi-partitioning scheme at the end of his third grade in school. The equi-partitioning scheme is essentially a scheme that some children use in breaking off one of several imagined parts from a "continuous" item. My construction of this scheme occurred in a situation where we asked Jason and the child he was working with at that time, Patricia, to cut a piece of candy off from a "candy stick" for one of four people. The children were using a computer tool TIMA: Sticks[2] that we had designed for the teaching experiment. In TIMA: Sticks, a segment can be drawn using the mouse cursor, the segment can be marked using hash marks, marked parts can be pulled out of the whole stick (an operation that left the marked stick intact), copies of the pulled part can be made, and the copies can be joined together to make another stick. To make the share of one of four people, Jason independently marked off one part of a segment they had drawn in the screen and broke the part off from the marked segment. In a test to find if the part was one of four equal parts, Jason made three copies of the part and joined them together with the pulled part. If the segment made by joining the parts together was shorter or longer

---

[2]TIMA: Tools for interactive mathematical activity.

than the original segment, he independently made a new estimate and proceeded to test it. He continued on in this way until they made one of four equal parts. Marking a segment once in estimating one of four equal parts implies that he used his concept, 4, in gauging where to make the mark, which means that he used 4 as a partitioning template. The ability to iterate the pulled part in a test to find if the estimate was one of four equal parts was inherited from the iterability of his units of 1. These two ways of using his concept of 4 was introduced by him in his interactions in TIMA: Sticks, and it served him in constructing the equi-partitioning scheme. The purpose of the equi-partitioning scheme is to estimate one of several equal parts of some quantity and to iterate the part in a test to find whether a sufficient number of iterations produce a quantity equal to the original.

An Account of Laura's Simultaneous Partitioning Scheme

I cannot emphasize enough that Jason's equi-partitioning scheme was established using his numerical concept, 4. To illustrate how Laura used her numerical concepts through 10 in partitioning, I select a protocol from the teaching episode held on December 7 of her fourth grade in school when she was working with Jason.

Protocol I: Drawing a Stick That is 1/10 of Another Stick.
> T: Can each one of you draw 1/10 of that stick? The one who wins will be the one that will be closer.
> L: (Draws her estimate) Right there!
> J: (Looks at the screen for some time and draws his estimate)
> L: That's the same!
> J: No it isn't, no it isn't!
> L: Ok! I will go first here! (She REPEATs[3] her estimated stick to make a 10-part stick, and it is too long.)
> J: (Even though his estimate is longer that Laura's, he still REPEATs it to make a 10-part stick to check.) Oh gosh! (Both children giggle.)
> T: You want to try one more?
> L: I want to try it one more time!
> T: 1/10, all right!
> L: 1/10, 1/10! Ok! This is my color! Ok that was too long... ok! That long! (Draws her estimate.)
> J: (Draws his estimate; both children laugh.)
> L: (REPEATs her estimate reciting the times out loud, and the estimate is very accurate.) Just about!

---

[3]The computer action, REPEAT, makes a copy of the repeated stick and joins it to what precedes.

T: Very close!! Let's see Jason. That's very nice!
T: (Speaking to Jason) What do you think yours is? Too short or too long?
L: Too short!
J: (REPEATs his estimate, which is shorter than Laura's and so produces a shorter stick than the unit stick. The children giggle.)
T: All right a little bit too short!

The children became deeply engaged in the task and expressed pleasure at making an estimate by drawing a stick and then testing their estimates by using REPEAT. The initial estimates of both children were closer to 1/8 and 1/7 of the unit stick (Laura and Jason, respectively) than to 1/10 of the unit stick, and Laura's second estimate was uncannily accurate. Both of the children used the iterative aspect of the units of their numerical concept, 10, to test their estimates by iterating them 10 times and comparing the results against the original stick.

That Laura made such an uncannily accurate estimate on her second trial may have been fortuitous. So, in the next teaching episode, held on February 8 of the next calendar year of the children's fourth grade, the teacher posed a task involving sharing a stick into eight equal parts. The reason for posing this task was to investigate two things. First, making an estimate of 1 of 10 parts of a stick would seem to involve the child using his or her numerical concept, 10, to project units into the stick. So, the task of Protocol II was presented to corroborate that the children used their numerical concepts in partitioning. The teacher also wanted to explore whether Laura's use of iteration in Protocol I was specific to drawing an estimate. The initial task was, after Jason drew a stick the same length as a Snickers candy bar, to share the candy bar equally among eight people. The teacher imposed the constraint that they could make only one mark on the stick to encourage the children's use of iteration to actually produce the eight parts.

Protocol II: Sharing a Candy Bar Among Eight People by Making Only One Mark.
   T: (Counts all the persons in the room aloud: 1, 2, 3, 4, 5, 6, 7, 8.) Your first task is to share this candy bar among these people. But use only marks. Remember you can move marks. But mark only the share of one person and use that to create all the shares. Just one mark for eight people. Go ahead.
   J: (To Laura) go ahead.
   L: (Takes the mouse and activates MARKS) But we can use a lot of marks to. ...
   T: Use one mark, if it will not come out as a fair share then you can use another, but try to make it as close as you can in the beginning.
   L: (Activates MARKS again and tries to estimate where to put the first mark. She makes an uncannily accurate estimate. The mark she makes on the stick is apparently 1/8 of the unit stick, but she is yet to produce the remaining parts.)

T: You know, we can still play with the screen. Remember PULL PARTS[4] and REPEAT? (Indicating to Laura that she isn't done).

L: Ok, there's....

T: You remember, PULL PARTS and REPEAT.

L: So, can I make another mark?

T: No, no, just one mark. Now see if it's a fair share.

L: (Seems confused and looks for an action to use in TIMA: Sticks.)

T: Do you want to pull the part first?

L: Okay. (Activates PULL PARTS and pulls the greater of the two parts from the marked stick. She sets the cursor over the smaller piece) Do I do this piece, too (the marked piece that she estimated as the share of one person)?

T: Which one do you want to use to check to see if it's 1/8?

L: Umm, this one (points to the 7/8-stick she pulled out from the marked unit stick)?

T: (To Jason) Jason, do you have an idea?

J: (Nods yes and takes the mouse, drags the 7/8-stick to the top of the screen, then pulls Laura's estimate from the marked stick).

T: Can you tell Laura what you are going to do?

J: I'm gonna' ... pull one of these (points to the estimated 1/8 part of the marked unit stick) and put it under there and see if. ...

L: (Enters Jason's talk, nodding yes) and REPEAT.

T: Okay.

J: (REPEATS the 1/8-stick until the end of the repeated stick reaches the end of the original stick. The resulting 8/8-stick seems to be exactly the same length as the unit stick.)

T: Wow, Wow, Wow! Laura you made it so quickly!! 1, 2, 3, 4, 5, 6, 7, 8. I don't believe it! Isn't that great! (The children then marked the unmarked original stick using the 8/8-stick as a template.)

The accuracy of Laura's estimates in Protocols I and II should not be regarded as fortuitous. Her estimate of 1/10 in Protocol I was uncannily accurate on her second trial and there were other occasions where she made similar accurate estimates. Her comment in Protocol II, "But we can use a lot of marks to..." should be considered as indicating that she visualized marks on the stick so that eight parts would be formed. She could then accurately gauge the length of one of the parts.

Although Laura made an uncannily accurate estimate of 1/8 of the stick, she did not independently use PULL PARTS and REPEAT to produce the other parts and to justify why her part was one of eight equal shares. Laura definitely could use her number concepts up to 10 as templates for partitioning blank sticks in the true sense of a partitioning. In fact, in Protocol I, it is plausible that she used the composite unit, 10, as a

---

[4]The computer action, PULL PARTS, makes a copy of a designated part of a marked stick that is superimposed on the part and that can be dragged off from the part.

partitioning template in making her estimate. In that case, she would simultaneously project the units of her composite unit into the blank stick and experience the parts as co-occurring. In Protocol II, this partitioning activity seemed to exclude any need to verify the part she marked off by pulling the part from the original stick and iterating it eight times to make a test stick. The operation of iteration unquestionably was available to her as indicated in Protocol I. But, in that case her estimate was not a part of the stick of which she was estimating a part, whereas in Protocol II, her estimate was a part of the original stick. Although iterating, partitioning, and disembedding were operations that were available to her, she seemed to use only partitioning in Protocol II. That is, she used her number concept, 8, to project units into the stick but not to iterate the part she marked off to produce an 8-part stick to compare with the original stick. Jason, on the other hand, disembedded the part of the stick Laura had made and iterated that part in an attempt to find if it was indeed 1/8 of the unit stick. In that Laura made such uncannily accurate estimates when partitioning a blank stick into up to 10 parts, I refer to the scheme she used as a *simultaneous partitioning scheme* to distinguish it from Jason's equi-partitioning scheme. She could engage in simultaneous partitioning activity, as did Jason, and could engage in iterating activity as did Jason, but she did not independently use iterating to find if an estimated part of a stick could be used to reconstitute the stick as did Jason. For this reason, partitioning and iterating did not seem to be parts of the same psychological structure for her as they were for Jason.

THE PARTITIONING SCHEMES AND COMMENSURATE FRACTIONS: AN INITIAL PREDICTION

Given the differences in the equi-partitioning and the simultaneous partitioning schemes, I hypothesize that these differences would be manifest in the schemes the children would construct to produce fractions commensurate with a given fraction. The rationale for this hypothesis is contained in what I have called the *splitting operation* (Steffe, 2002). When asked to make a stick so that a given stick is, say, five times longer than the one to be made, the child needs to posit a hypothetical stick such that repeating that stick five times would be the same length as the one given. That is, the child would need to not only posit a hypothetical stick, but also to posit the hypothetical stick as one of five equal parts of the given stick and to understand a priori that the partitioning of the given stick and the iterating of the hypothetical stick produce the self-same partitioned stick. This constitutes a composition of partitioning and iterating.

In the case of the equi-partitioning scheme, partitioning and iterating are yet to be composed. The results of a mental partition produces a situation the child uses to estimate an actual part of the whole stick. The child then uses this estimated part in iteration to produce a partitioned whole that is equivalent rather than identical to the original. This is quite different than producing a hypothetical stick such that repeating the stick, say, four times, would produce a stick identical to the given stick. During their fourth grade in school, neither of Jason nor Laura constructed the splitting operation in spite of our attempts to encourage it (Steffe, 2002). Still, given Jason's equi-partitioning scheme, I predicted that splitting should emerge in his case. I was uncertain about whether splitting would emerge in Laura's case. My prediction was that her lack of disembedding an estimated part of a stick and using it in iteration to produce a partitioned stick would prove to be an internal constraint in her construction of the splitting operation

At the outset of the teaching experiment with Jason and Laura while they were in the fifth grade, my prediction was that the child's ability to split a stick would be essential in the production of a fraction commensurate with a given fraction. But at that time I had not constructed the operations that children use to produce a fraction commensurate with a given fraction. The goal of the teacher-researcher was to bring forth the productive thinking in the children necessary to transform a given fraction into a commensurate fraction, and my goal was to analyze whatever mathematical activity the teacher-researcher was able to bring forth in the two children and, in this context, generate a learning trajectory of the two children as they engaged in activities designed to bring forth commensurate fractions.

THE EMERGENCE OF A FRACTIONAL MULTIPLYING SCHEME IN JASON

The initial teaching episode with Jason and Laura while they were in the fifth grade occurred on October 25. The primary goal of the teaching episode was for the children to reestablish the fractional schemes they had constructed during the fourth grade in the context of using TIMA: Sticks. During fourth grade, Jason had constructed a scheme for making fractions that I regard as partitive by a modification of his equi-partitioning scheme (Tzur, 1999). I call a fractional scheme a *partitive fractional scheme* to emphasize that the dominant purpose of the scheme is to partition a continuous unit into a specific number of parts, take one part out of those parts and establish a one-to-many relation between the part and the partitioned whole, and then iterate the part enough times to produce a fraction, say, 3/7.[5] The iterative aspect of the scheme can be used in justifying or verifying that a unit part of a continuous unit is one of so many equal parts and that the unit part, when iterated, can be used to produce the measure of the continuous unit. Jason knew, for example, that one of ten parts of a stick was 1/10 of the stick, and that because the whole stick is 10 little pieces, "it is a how long the whole stick is. So one whole stick is 10 pieces of those little ones." Understanding that 10 of a 1/10-stick consti-

---

[5]The partitive fractional scheme cannot be used by a child to produce fractions beyond the fractional whole.

tutes the length of the whole unit stick is basic, and I regard it as essential for a scheme to be called a partitive fractional scheme.

Laura did not seem to understand that 10 tenths constituted the length of the whole stick even though she did say "10 tenths" for the result of repeating a 1/10-stick 10 times. This separation of the results of iterating and partitioning continued throughout her fourth grade, and it reappeared in the teaching experiment in the fifth grade. For this reason, I call the fractional scheme she constructed using her simultaneous partitioning scheme a *part-whole fractional scheme* to distinguish it from Jason's partitive fractional scheme. For Laura, a fraction like 7/10 meant to take seven parts out of 10 parts and then to make a 7-to-10 comparison. For Jason, it also referred to the length of a stick that he produced by iterating a 1/10-stick seven times.

Recursive Partitioning and the Fractional Composition Scheme

While the children were in the context of showing the teacher all they knew about 3/4 in the October 25 teaching episode, an entirely unplanned event occurred that led to my construction of the operation of recursive partitioning and the factional composition scheme.

Protocol III: Making a Fraction of a Fraction.
J: (Makes a copy of a 4/4-stick,[6] which he had been using, and colors three parts of it. He then pulls these three colored parts out of the 4/4-stick, releases the mouse, and sits back in his chair.)
T: Ok, so now you have 3/4. So, now I want to find another way to deal with PULL PARTS and make 3/4.
L: You can make it smaller!
T: Go ahead. I don't see what you mean so let's see.
L: (Takes the mouse and pulls one part from the 4/4-stick.)
T: Now, can you use that one to make 3/4?
L: (Dials PARTS to 4 and clicks on the pulled part. Following this, she colors three of the four parts and uses PULL PARTS to pull them from the stick.)
T: Wait, wait, wait, wait. Now I want to ask you a question because what you did was so nice! Can you give a name, a fractional name, can you tell me how much this is out of the whole (spans the unmarked original whole with his index finger and thumb)?
L: 3/10.

---

[6]A 4/4-stick was a stick marked into four equal parts. To be called a 4/4-stick, the four-part stick must be interpreted in that way by the child.

J: (Puts hand under his chin and thinks) 3/16.

T: Because you have different answers and you are a team, you want to give me one answer and explain to each other until you get to a solution.

L: Oh oh! (Meaning, "we are in trouble") well, we had ... I don't know...

J: (Points at the 3/16-stick Laura made) See, if we would have had it in that (points to each part of the 3/4-stick he made by pulling parts) 4, 4, 4, and 4—16. But you colored 3, so it is 3/16!

L: Oh! I thought you meant the thing we first started with was a 10 (presumably referring to the 4/4-stick as a 1/10-stick).

After Laura pulled one part from the 4/4-stick, the teacher intervened and asked Laura if she could use that part to make 3/4. His expectation was that Laura would use REPEAT or COPY to make a stick that was 3/4 of the unit stick. It was a complete surprise to the teacher that she made 3/4 of the 1/4-stick. In retrospect, her making 3/4 in this way was a confirmation of the lack of the operation of iteration in her part-whole fractional scheme.

The teacher seized upon the opportunity and asked the children to give him a fractional name: "Can you tell me how much this is out of the whole?" Laura's answer of "3/10" and her explanation ("I thought you meant the thing we first started with was a 10") had no observable basis in the context of the teaching episode, as the children had not made 1/10. Jason's answer of "3/16" and his explanation of his answer was also a complete surprise. It was as if he had constructed new operations in his partitive fractional scheme over the summer vacation.

Although it was Laura who made 3/4 of 1/4 in Protocol III, Jason's actions indicate that he regarded her actions as if they were his own. What this means is that he mentally performed the actions that he observed Laura carry out. So, when the teacher asked, "Can you tell me how much this is out of the whole," Jason had already produced the 3/4 of 1/4 with Laura. Jason's comment ("See, if we would have had it in that (points to each part of the 3/4-stick he made by pulling parts) 4, 4, 4 and 4—16") indicates a kind of reversibility of the composition of two particular instantiations of his fractional scheme—make 1/4 and then make 3/4 of 1/4 and posit how much the 3/4 of 1/4 is of the whole stick. But it goes beyond what I consider as a reversible partitive fractional scheme that would permit a child who is given, say, 7/10 of a fractional whole to make the fractional whole by reversing the operations involved in making 7/10 of the fractional whole.

In a teaching episode held on May 12 of his fourth grade, I inferred that Jason constructed reversibility of his partitive fractional scheme. He was given a blank stick and was asked to make the whole stick if the blank stick was 7/10 of the stick. Jason partitioned the blank stick into seven parts, pulled three parts, and joined them to the seven. Not only did this involve reversing the sequence of operations, it also involved reversing the operations. When making a 7/10-stick, the last operation involved is to conceptually unite seven individual parts together into a composite unit containing seven elements. During the May 12 teaching episode, Jason performed the inverse of this uniting operation by partitioning (disuniting) the blank 7/10-stick into seven equal parts using PARTS. That he regarded each part as 1/10 of some stick of which the 7/10-stick was a part is solidly indicated by his pulling three parts out of the 7/10-stick and joining them to the 7/10-stick using JOIN. This way of making the 10/10-stick given a 7/10-stick can be regarded as the reverse of making a 7/10-stick given a unit stick. Laura too provided indication of reversibility of her part-whole fractional scheme.

Given 3/4 of the 1/4 Laura made, the expected results of a reversible partitive (or part-whole) fractional scheme would be to produce a partition of the whole fractional stick whose elements are of the same size as the elements of the 3/4 of the 1/4. Of course, the discrepancy between the whole stick not being partitioned into these elements and the expectation that it be partitioned in that particular way constitutes the perturbation that drives the reversible scheme's activity. But the operations of a reversible partitive fractional scheme are not sufficient to eliminate this perturbation in the case of 3/4 of 1/4. The only way to partition the whole stick in this case is to distribute the partitioning of 1/4 of the stick across the three remaining fourths as did Jason. So, a child who has constructed only a reversible partitive fractional scheme is left in a search mode induced by the perturbation with no action to perform.

Jason made a modification in his reversible partitive fractional scheme that I call *recursive partitioning*. Recursive partitioning is quite different than directly forming the goal of partitioning a stick into four parts and then each part into four parts. This latter case would involve partitioning a partition, but I would not consider it necessarily as a recursive partitioning.

For a composition of two partitionings to be judged as *recursive* in a fractional context, there must be good reason to believe that the child, given a partial result of the composition (like 3/4 of 1/4), can produce the numerosity of the full result. But this is not all, because the child must also use the second of the two partitions (the one that is not fully implemented) in the service of a goal different from simply intending to produce a partition of a partition. Jason's goal was to find how much 3/4 of 1/4 was of the original stick, which is quite different than the goal of partitioning a partition. The importance of this analysis of goals is that the composition of the two partitions is called in the service of the goal of the reversible partitive fractional scheme. This amounts to embedding the composition of the two partitionings as a subscheme in the reversible partitive fractional scheme.

If Jason's modification of his reversible partitive fractional scheme proves to be more or less permanent, then it can be said to be a *functional metamorphic accommodation* of that scheme. A functional accommodation of a scheme occurs in the context of using the scheme. To be metamorphic, the accommodation must occur independently, reconstitute the scheme on a new level, and reorganize the scheme at that level (Steffe & Wiegel, 1994, p. 126). Later in the discussion, I explain why I regard the modification that Jason made in his scheme as metamorphic. Here, I want to emphasize that I think of the reorganized scheme as a *fractional composition scheme*, which is a fractional multiplying scheme. The goal of this scheme is to find how much a fraction of a fraction is of a fractional whole, and the situation is the result of taking a fractional part out of a fractional part of the fractional whole, hence the name *composition*. The activity of the scheme is the reverse of the operations that produced the fraction of a fraction, with the important addition of the subscheme, recursive partitioning. The result of the scheme is the fractional part of the whole constituted by the fraction of a fraction.

An Apparent Recursive Partitioning

Given the results of Protocol III, the teacher now had two goals. The first was to explore whether the change experienced in Jason's way of operating with fractions recurred in other situations, and the second was to bring forth a fractional composition scheme in Laura as a result of interacting mathematically with Jason. Laura's comment, "Oh! I thought you meant the thing we first started with was a 10," indicates that she did in fact return to the original stick and make an estimate of how many little pieces would be in that stick. This would constitute a fulfillment of the expectation of a reversible part-whole fractional scheme but without implementing its activity. Instead, it would be an estimate of how many little pieces were in the whole unmarked stick. Based on her simultaneous partitioning scheme, we know that she had a propensity for making such estimates.

Protocol III: (First Continued)
  T: (To Laura) Ok, Laura, so what did you do? You said this is 3/4 of that small one (pointing to the 3/16-stick and to the 4/16-stick, respectively). This is so nice. Show me more about 3/4.
  J: (While the teacher is talking to Laura, Jason pulls two parts from a copy of the original 4/4-stick he made, erases the mark separating the two parts of this stick, partitions this 1/2-stick into four parts using PARTS, and fills three of these four parts with a different color.)
  L: (Looks at the results of Jason's activity) Oh, I know another way.
  T: (To Laura) Wait, wait, wait, wait, wait, don't try. That's good, keep it (referring to the result of Jason's activity).
  J: I got two of these, erased that line, and I ... um ... put four pieces.
  T: How much is this one (the 3/4 of 1/2 that Jason made) of this one (the unmarked unit stick)?
  J: (After approximately 10 s) 3/8 (confirming his construction of a fractional composition scheme)!

Jason, saying "3/8," provides corroboration that recursive partitioning did indeed recur in his independently executed productive activity of making 3/4 of 1/2 of the original 4/4-stick he made. Changing the situation from making 3/4 of 1/4 to making 3/4 of 1/2 does indicate that he could willfully generate situations of his fractional composition scheme. However, he did not seem to know in advance that he had made a 4/8-stick because it took him approximately 10 s to say "4/8" after the teacher asked, "How much is this one of this one?" Nevertheless, by changing the situation he seemed to have the confidence that he could produce a fractional number word for the stick he made, which is solid indication that his fractional composition scheme was an anticipatory scheme. Laura initially could not say how much the stick that Jason made was of the whole unit stick, so the teacher continued exploring how she might respond to the situation.

Protocol III: (Second Continued)
T: Ok (Looks at Laura in anticipation of her answer).
L: (Sits quietly for approximately 18 s) I don't know, … am … am … am.
J: (Imitates Laura in a teasing manner) am. …
L: (Laughs at Jason's teasing).
T: Do you want Jason to explain to you because you work as a team.…
L: I don't know. I guess that's right because that came out of there (pointing to the 4/16-stick she made originally and to the 4/4-stick she pulled it from); those two (pointing to the 4/8-stick Jason made as if it was still broken into two rather than four parts) came out of … (pointing to the 4/4-stick).
J: (Interrupting Laura's analogical reasoning) I got two out of this one (pointing to two parts of the 4/8-stick), and then we have four of these (pointing to the 4/4-stick), and two and two, … (meaning there are four 2s) and there's four (pointing at the 4/8-stick he made).
T: (To Laura) Because I am not sure that I understand Jason right, I want you to explain to me what he said.
L: These two (points to each 1/2 of the 4/8-stick Jason made with two fingers, one finger on each half) came out of here (points to two 1/4 of the 4/4-stick with two fingers, one finger on each 1/4), and that would be 2, 4, 16 (counting 2, 4 apparently referred to counting 1/8s, and 16 apparently referred to putting four parts into each 1/4 of the 4/4-stick).
T: But, Jason said 3/8.
L: Oh, well, … Oh, well, … Oh, Ok. See, four and four (pointing to the first two parts of the 4/4-stick, again using 4 rather than 2 in partitioning) and then four (pointing to the 4/8-stick), and then these three (pointing at the 3/8-stick Jason made) came out of here (the 4/8-stick).
J: (Makes another explanation.)
L: Ok. (Creates a new situation. She draws a new stick approximately the same length as a 1/8-stick, partitions it into four parts, and pulls three of the four parts in a reenactment of both her and Jason's making of three out of four parts.)

J: Oh, my God!!

T: Before you try to find out how much this three is of this one (the un-marked stick in the RULER), what is this 3/4 of?

L: (While the teacher is speaking, Laura nods her head toward the screen and subvocally utters number words) That's 4/32!!

Laura made an analogy between her making 3/4 of 1/4 and Jason making 3/4 of 1/2 before Jason interrupted her in an explanation of why the stick he made was a 3/8-stick. The apparent analogy indicated an awareness of a similarity in operating. Starting from the point where Laura made the analogy, she made rapid progress to explicitly counting by 4s in distributing partitioning each of the eight parts into four parts (the last two lines of the protocol). Whether this distribution of partitioning constituted recursive partitioning as explained in Jason's case is at issue because her goal was to find how many parts would be in the 8/8-stick if she were to partition each 1/8 into four parts rather than to find how much the three parts were of the 8/8-stick. This is indicated by her answer "4/32" rather than "3/32." Given the interaction between the two children, Jason's explanation seemed to be the reason why she was able to initiate partitioning a partition in the situation. Whether she embedded a recursive partitioning scheme in her reversible part-whole fractional scheme awaits further investigation.

The alternative is that she abstracted using her part-whole fractional scheme as a result of making 3/4 of 1/4 and making 3/4 of 1/2. Following upon two such sequential applications of her part-whole fractional scheme, the perturbation induced by her unfulfilled expectation of finding how much of the fractional whole was produced by Jason making 3/4 of 1/2 focused her on listening to Jason's explanation. She could recognize his explanation because she, too, had established a units-coordinating (or multiplying) scheme.[7] In this case, her choice of using her units-coordinating scheme would not be made independently, but as a result of observing Jason using his scheme, and the relation she established between her units-coordinating scheme and her reversible part-whole fractional would be an association rather than an embedding.

So, rather than constructing a fractional composition scheme, she may have constructed an associative chain of schemes, in which any scheme in the chain was triggered by the results of the scheme immediately preceding. If so, then the major difference between her and Jason's schemes would be that Jason could independently use his scheme whereas Laura would be unable to engage in the productive thinking necessary to produce a result of her chain of schemes in the absence of some sensory situation that triggered each scheme of the chain.

---

[7]In a units-coordinating scheme, given two composite units, say 2 and 4, the child distributes the unit of 2 across the units of the composite unit of 4, producing a sequence of 4 units of 2.

Commensurate Fractions

Establishing 4/12 as 1/3.   To further explore the nature of Laura's apparent progress, I turn now to exploring whether she could engage in the operations necessary to establish 4/12 as 1/3. In a teaching episode held on November 8, the teacher's plan was to begin by asking the children to partition a stick into 12 parts but under the constraint that they could dial PARTS up to 12 but not include 12.

Protocol IV: Sharing a Cake Among 12 People by Sharing Shares.
- T: Copy a cake and divide it among 12 kids (there was a blank stick on the screen that the children pretended was a cake and could copy).
- J: (Grabs the mouse and makes a copy of the stick. He then dials PARTS to 12.)
- T: Can you do that? ... Don't cheat on me!!
- L: Can we use PARTS at all?
- T: Sure, you can use PARTS, PULL PARTS, everything, but ah. ...
- L: (Dials PARTS to 11 and clicks on the stick. She then counts the parts, pointing to each part with the mouse cursor) 1, 2, 3, 4, 5, 6. ...
- J: (Interrupts Laura) 11?
- L: (Uses PULL PARTS to pull the last part out of her 11-part stick and joins it to the end of her stick.)
- T: Now you made another cake that we don't have. This is the cake (pointing to the original stick). We would like to add to our cake, but we can't!

Laura's action of partitioning the stick into eleven parts and then pulling the last part out and joining it to the 11-part stick is certainly rational, and it served her in making a 12-part stick that could be shared equally among 12 children. But she did not engage in recursive partitioning. Not being allowed to dial PARTS to 12 served as a constraint in her direct use of PARTS. This situation was as close as we could come to establishing a situation that would induce recursive partitioning without stipulating how she should operate. Jason's recursive partitioning operation had been activated, however, and he was explicitly aware that what Laura did was not what he wanted to do.

Protocol IV: (Continued)
- J: (Grabs the mouse while the teacher is talking) Heyyy...! (Drags the stick Laura made into the TRASH and makes another copy of the stick. He then dials PARTS to 3 and clicks on the RULER, then dials PARTS to 4 and clicks on each of the three parts of the 3/3-stick, making a 12/12-stick.)
- T: Can you explain to me as a child in this party why we would now have the same, an even piece for each one?
- L: He had three pieces and he added four in each thing.
- T: What makes it even? Tell me more because I am not sure I understand. Sounds good to me, but I'm not sure that. ... Can you pull the piece of one kid out?

L: (Takes the mouse and pulls the first part of the 12/12-stick out from the whole stick.)

T: Can you tell me what fractional part of the cake is that one?

L: (Immediately) 1/12!

J: (Following Laura) 1/12.

T: Can you pull out, ... four pieces?

J: (After dragging the 1/12-stick into the TRASH, pulls four parts from the stick using PULL PARTS.)

T: How much of the whole cake is the share of the four kids?

J: (Immediately) 4/12.

T: Can you explain that to me?

J: There's 12 pieces and there are four that's colored (Jason had colored the four pieces he pulled out).

L: (Repeats what Jason said) 12 pieces and four colored.

T: Ok, now we are coming to a problem. Can you measure it and see if what we have is 4/12?

L: (Takes the mouse and uses MEASURE[8] to measure the 4/12-stick. While she is measuring, she says "1/12" as a guess of what will appear. Jason guesses "3/4." 1/3 appears in the NUMBER BOX.)

J&L: (Both are surprised that 1/3 appeared) 1/3!? (Almost simultaneously)

J: (Almost immediately) Oh, oh, I see that!! (Grabs the mouse)

L: (As Jason grabs the mouse) Oh, I do too because first he, first he (pointing to the 12/12-stick excitedly) ... first, he put it in three pieces and he made four in each thing (pointing three times at the 12/12-stick from the left to the right in designation of "thing") so four, that would be one of the four (pointing back and forth between the 12/12-stick and the 4/12-stick), one of the four pieces, ah, that would be one of the three pieces (making brackets with her hands as if there is something in between them).

J: (As Laura finishes her explanation, activates REPEAT and repeats the 4/12-stick three times, making a 12/12-stick.)

T: Ok!! All right, that's very nice, so you did it, you gave me a very good explanation. You gave me one (points to Laura), and you gave me another one (points at Jason). (To Laura) did you see what Jason did?

L: No.

T: When you were talking, explaining to me and that was marvelous, he took the first piece, what did you do, Jason?

J: Repeated it.

T: How many times:

J: Two.

T: So, altogether, how many do you have?

---

[8]The MEASURE function of TIMA: Sticks provides a measure of the stick selected based on a stick that has been designated as the unit stick. The measure is given as a fraction in lowest terms.

J: Three. And that one was like, purple, and I put 4, 4, and 4, and that equals 12.

T: (To Laura) Which is exactly what you said, isn't it?

L: (Before the teacher addressed her, she was looking around the room, and gave no indication of understanding what Jason said other than a little head nod "yes.")

J: (Continues on explaining) And we colored, like we colored four, and its 1/3.

Jason's decision to partition the stick first into three parts and then each part into four parts is corroboration of the inference that he had constructed recursive partitioning operations. His manner of operating was distinctly multiplicative in that he seemed to be aware simultaneously of the result, 12, of his partitioning activity, and of 12 partitioned into three composite units each of numerosity 4. He seemed aware not only of this structure, but also of the operations that produce it—first partition the stick into three parts and then each of these parts into four parts.

Given Laura's lack of recursive partitioning operations, it was somewhat of a surprise that she explained why 1/3 came up in the NUMBER BOX after she measured the 4/12-stick using the stick in the RULER.[9] Her comments—"First, he put it in three pieces and he made four in each thing (pointing three times at the 12/12-stick from the left to the right in designation of 'thing') so four, that would be one of the four (pointing to the 4/12-stick), one of the four pieces, ah, that would be one of the three pieces"—do indicate that she established three composite units with four elements in each in her regeneration of Jason's partitioning activity, and that she regarded the piece Jason made as one of these three composite units.

When 1/3 appeared in the NUMBER BOX, it was surprising to her and it is no exaggeration to say that she experienced perturbation. We know that for Laura, 1/3 meant to partition a stick into three equal parts and then disembed one part from the three parts and compare the part to the whole. I assume that 1/3 activated this scheme of operations and she used it in regenerating Jason's actions. Her saying "First he put it in three pieces. ..." indicates that she regenerated Jason's act of partitioning the stick into three parts using the partitioning operations in her concept of one third. Further, her comment "he made four in each thing (pointing three times at the 12/12-stick from the left to the right in designation of 'thing')" indicates that she regenerated Jason's actions of distributing partitioning into four parts across the now given three abstracted unit items. This seemingly produced a novelty in her concept of 1/3 in that she now regarded the 12/12-stick as composed of three composite units with four elements in each unit as indicated by her making brackets with her hands as if something was between them. So, not only did she regenerate her experience of Jason making the second partition, she also united each of the four elements produced into a composite unit, an operation that she was quite capable of performing. Finally, her

---

[9]The RULER is a designated strip of the computer screen into which a stick can be copied, thus designating it as the unit stick for measurement purposes using the MEASURE function.

comment "that would be one of the four (pointing back and forth between the 12/12-stick and the 4/12-stick), one of the four pieces, ah, that would be one of the three pieces" indicates that she disembedded one of the composite units from the three others and compared it to the others.

So, there were two novelties in Laura's use of her concept, 1/3. First, she inserted units of four into each of the three units she produced when regenerating Jason's partitioning the stick into three pieces and reasoned as if these composite units were units of one. Second, she used her concept of 1/3 in monitoring her activity. What monitoring means in this case is that she seemed explicitly aware of three composite units of four and, further, that she was aware of focusing her attention on the four unit items within each of the three composite units and then shifting her attention from these four items to the three composite units. She was aware of two levels of units, and could deliberately shift her attention between the two levels as it suited her purpose. But this does not complete the account of her monitoring, because she used her concept of 1/3 to structure the situation as 1/3, and in doing so, she made checks to be sure her operations with composite units were also operations for making 1/3. That is, her actions of regenerating Jason's actions fed back into her concept of 1/3 in the process of regenerating them. If Laura's ability to monitor her making of 1/3 was a permanent modification, then she should be able to engage in productive thought in future situations like the one of Protocol V.

Jason's repeating of the 4/12-stick also indicates that he used his concept of 1/3 in reconstituting the 4/12-stick as one of three composite units. This is the first time that Jason seemed to be explicitly aware that if a composite part can be repeated three times to reconstitute the whole composite unit, then the composite part is 1/3 of the whole. He seemed aware before he repeated the 4/12-stick to make the 12/12-stick that repeating the former three times could produce the latter. This is indicated by his actions of repeating the 4/12-stick as well as by his answer, "Three. And that one was like, purple, and I put 4, 4, 4, and that equals 12," after the teacher asked, "So, altogether, how many do you have?" He used this result in making his judgment that the 4/12-stick was indeed 1/3, as indicated by his comment, "and we colored, like we colored four, and its 1/3."

If Jason was aware before he repeated the 4/12-stick to make the 12/12-stick that the latter could be produced by repeating the former three times, then this would indicate that he could take a unit of units of units as a given without needing to actually engage in the operations that produce this structure. If he had indeed constructed the ability of taking the results of actually making a unit of units of units as a given, this would constitute a major advancement in his numerical operations, because he could now use this ability in producing an equi-partitioning scheme for composite units. Whether Laura could take the structure of a unit of units of units as a given prior to operating is also at issue. The importance of children independently producing a fraction commensurate to a given fraction cannot be underestimated to make the claim that they have constructed a commensurate fraction scheme. Both children could assemble the operations that were needed to justify why a 4/12-stick can be also a 1/3-stick given the computer read out of 1/3 after measuring the 4/12-stick. The difference in the two children's justifications is that Jason used REPEAT to produce a three-part stick, thereby indicating that he was in the process of establishing an equi-partitioning scheme for composite units that contained the same operations as his equi-partitioning scheme for continuous units of 1. Laura, on the other hand, mentally reorganized the perceptual material that was available to her into three composite units each of numerosity 4 in a reenactment of Jason's actions. Moreover, Jason's explanations of repeating the 4/12-stick three times seemed to carry little significance for Laura. There was no indication that she assimilated Jason's iterative actions and his explanation of them, which is compatible with the assumption that iteration was not an operation of her fractional scheme. Given the differences, the expectation was that Laura would be restricted to regenerating actions such as those carried out by Jason in establishing commensurate fractions, whereas Jason would be able to generate a fraction commensurate to a given fraction as a result of productive thinking.

Establishing 3/15 as 1/5 and 5/15 as 1/3.   In the teaching episode held on November 15, two occasions arose in which it is possible to make a judgment concerning the operations Laura used to justify why 4/12 is commensurate with 1/3. The teacher decided to use 15 as the number of kids that came to the party.

Protocol V: Three Fifteenths and One Fifth.
   T: Fifteen kids came to the party. Start with three cuts on the birthday cake.
   L: (Takes the mouse and dials PARTS to 3 and clicks on a copy of the sick in the RULER. She then colors the parts each a different color, dials PARTS to 5 and clicks on each of the three parts.)
   T: (To Jason) do you agree with that?
   J: (Shakes his head "yes.")
   T: Ok, that seems to be all right, but I want to check it (takes the mouse and begins to proceed).
   L: Oh no!!
   T: But I want to check it!! (Then indicates to Jason that he is to pull one part to check it.)
   J: (Pulls the first of the 15 parts from the 15/15-stick that Laura made.)
   L: It's going to be one, one, one. ...
   J: Fifteenth!
   T: 1/15, you say?
   J: (While the teacher is talking, measures the 1/15-stick and 1/15 appears in the NUMBER BOX.)
   T: (Laura draws his attention to the NUMBER BOX.) All right, so you are okay. (On suggestion by a witness) What would be the share for three people (changing the fractional part so it would not correspond to the way the 15 parts of the stick were made)?
   L: (Immediately) 3/15!

J: (In agreement with Laura) 3/15. (He then makes a 3/15-stick by repeating the 1/15-stick.)

T: I say it would be 1/5!

J: (Looks intently at the sticks in the computer screen) I agree.

L: (Sits and looks intently at the sticks in the computer screen for approximately 15 s longer.) I don't agree.

T: (Asks Jason whether it is 3/15 or 1/5.)

J: Its 1/5.

T: It's 1/5?

L: I don't know.

T: (Asks the children to explain why it could be both.)

J: (Points to the 3/15-stick) There's five of these. ... (Takes the mouse and pulls a copy from the 3/15-stick and then makes copies of this copy. He then aligns four copies with the 3/15-stick directly beneath the 15/15-stick.)

L: I get it (in recognition of Jason's explanation)!

J: (Counts the copies of the 3/15-stick in the row, pointing to each in synchrony with uttering) 1, 2, 3, 4, 5. (Explains to Laura that these sticks go five times into the original 15/15-stick.)

T: (To Laura) you said it was 3/15, (To Jason) you said it was 1/5. It's both?

J: It's kind of both!

Presumably, when Laura said, "Its going to be 1, 1, 1...," she could have produced "fifteenths" in completion of the phrase because she had done so many times before with other unit parts. In fact, she did say, "3/15" for the share of three people. However, from that point on, she did not independently produce an explanation for why the 3/15-stick could also be a 1/5-stick. She had colored the three parts of the whole stick she made originally, so she had no visual cues that would provoke her units-coordinating scheme. Moreover, Jason had not yet offered an explanation, nor could she rely on how she made the 15 parts to justify why the share of three people would be 1/5. This was the first time she had to initiate operations on her own to explain why 3/15 and 1/5 were commensurate, and she was unable to do so. It is quite significant that after Jason aligned five 3/15-sticks end-to-end beneath the 15/15-stick in explanation for why a 3/15-stick could also be called "1/5", only then did Laura say, "I get it!" It wasn't the case that she was simply waiting for Jason to make an explanation, because she genuinely did not know why the teacher insisted on referring to the 3/15-stick as 1/5 (she said "I don't know" when pushed by the teacher). This observation corroborates the belief that she couldn't yet engage in recursive partitioning operations.[10] The monitoring activity

---

[10]Making such an explanation involves more than recursive partitioning in that it involves producing a composite unit (3) to partition each of the units of 5 to produce 15 smaller units. Nevertheless, recursive partitioning is involved in making the explanation because the child's goal is not the direct partitioning goal of partitioning the elements of a partition.

she engaged in when using her concept of 1/3 in reconstituting the 4/12-stick as a 1/3-stick in Protocol IV did not constitute an accommodation and occurred only because her units-coordinating scheme was called in the specific situation. Apparently, there is a major distinction to be made between Jason independently engaging in recursive partitioning to produce 12 pieces of cake before he made his explanation in Protocol IV and Laura regenerating the results of Jason's recursive operations when explaining why a 4/12-stick also could be called a 1/3-stick.

Protocol V: (Continued) Laura's Second Attempt to Justify Why a 3/15-Stick is Also a 1/5-Stick.

T: (Posing another task) Take a number of parts from up here (the original 15/15-stick). Fifteenths. A number of fifteenths. Take it out so here (the NUMBER BOX) we will get another number?

L: (Fills the first three parts of a 15/15-stick and then tries to measure them by measuring the whole 15/15-stick. However, 1 comes up in the NUMBER BOX, so she remeasures the whole stick several times obviously experiencing perturbation.)

T: You will have to pull a part out to measure it. You measured the whole thing.

L: (Pulls the first three parts from the 15/15-stick and measures them. 1/5 appears in the NUMBER BOX.) Oh (looking very disconcerted)!

J: I got it!! (Colors two more parts of the 15/15-stick and pulls them out, making a 5/15-stick. He then colors the remaining 10 parts of the 15/15-stick with two different colors, five one color and five another color. He then starts to measure the 5/15- stick.)

T: (Taking the mouse from Jason and laughing) Don't measure! Don't measure! What were you thinking about?

L: 3/15!

T: (To Jason) 3/15?

J: 1/3.

T: (Asks Jason to explain to Laura.)

J: (Points to the 1/3(5/15)-stick) there's five—three of these and that's just one.

L: So that's 5/3!!

J: (The teacher further pursued why the 5/15-stick could be both 5/15 and 1/3) Yes, but there are three groups, and one is fit into the other three!

The teacher's attempt to pose a task in which Laura would perform recursive partitioning failed. After he asked the children to take a number of fifteenths out of the 15/15-stick so they would get a different number in the NUMBER BOX, Laura interpreted his request using the 3/15-stick she had just experienced. When she finally measured it, and 1/5 appeared in the NUMBER BOX, she looked very disconcerted, indicating that 1/5 was not what she expected to appear. So, the situation in which she was involved simply did not activate the operations in which she engaged to say, "I get it" in the first part of Protocol V in that she did not use her concept of 1/5 to reconstitute 3/15 as 1/5. Jason, on the other hand, again engaged in independent mathematical activity, which corroborates the inference that he had constructed recursive partitioning operations and had made an accommodation in his partitive fractional scheme that permitted him to transform an appropriate fraction into a unit fraction.

Jason producing 1/3 for how much the 5/15-stick was of the unit stick also corroborates that he could iterate a composite unit of 5 to produce a composite unit containing three units of 5. After producing this unit structure, he was aware of three levels of units and understood that five out of the 15 individual units was commensurate to one composite unit of five out of three such composite units— "yes, but there are three groups, and one is fit into the other three!" In reply to why it could be 3/15 and 1/5 in Protocol V, he said, "It could be kind of both," indicating an equivalence and yet a distinction. So, I refer to the relation that Jason produced between 5/15 and 1/3 by using the term *commensurate* to stress his awareness that five individual units out of 15 such units could be transformed into one composite unit of five individual units out of three such composite units. I do not consider the relation he produced as a fractional equivalence, because that relation implies an abstracted relation that can be used to produce a class of fractions each equivalent to every other member of the class.

In the next task of the teaching episode, the teacher asked the children to choose how many kids could come to the party if the cake was cut into three parts to begin with. Jason said 27, and Laura agreed with him. In fact, after Jason said, "that's 3 times 9", Laura said, "I know," and stated her view of the relation as: "3 times what is 27?" Her statement of the relation indicates that she used her units-coordinating scheme to constitute the sharing situation after Jason suggested 27. It is a statement of anticipation of how many pieces each of the three parts of the cake should be cut into for there to be 27 pieces. However, when asked, "What would the share of three people be," she immediately said, "3/27." Jason, on the other hand, said, "3/9," and then corrected himself and said, "1/9." The teacher then asked the children to convince him that it is one of the two answers or whether it could be both answers. Laura argued that it would be 3/27 because there are three people and 27 pieces, which is a contraindication that she had constructed recursive partitioning. Jason argued for 1/9 and counted nine 3/27-sticks in the 27/27-stick in justification. He further measured a 3/27-stick. At this point, he seemed to be aware that it could be either of 3/27 or 1/9. In contrast to Jason's rather explicit use of his modified concept of 1/9 to reconstitute 3/27 as 1/9, Laura still fundamentally regarded the 3/27-stick as 3/27, not 1/9.

Producing Fractions Commensurate with 1/2

The difficulty Laura experienced in the November 15, 1993 teaching episode in explaining why a 3/15-stick could be also a 1/5-stick served as a constraint for the teacher. He realized that Laura did not engage in the operations necessary for generating an explanation regardless of his attempts to bring those operations forth. So, during the November 22 teaching episode, he decided to change the situations of learning to explore those accommodations that Laura might make in her part-whole fractional scheme that would enable her to transform a unit fractional part into a commensurate fractional part. Given that Laura had said, "3 times what is 27?" in formulating a way to find how to cut a cake into 27 pieces that had already been cut into three pieces, the teacher decided to start with one half of a cake and ask the children to take a piece of the cake that would be the same size as one half of the cake but which was a different fraction. He posed this task because Laura could posit an unknown number of parts into which each of three parts could be partitioned. Protocol VI picks the children up in the first task of the teaching episode in which Laura had already partitioned a stick into two parts, pulled one part, and labeled it "1/2." Most of the sticks mentioned in Protocol VI were copies of a unit stick which the children pretended was the uncut cake.

Protocol VI: Producing Fractions Commensurate with 1/2.
- T: What will be the next one that will be one half of the cake, but another fraction?
- L: What am I supposed to do? (She then drags a copy of the unit stick that is positioned at the top of the screen beneath the 1/2-stick. The 1/2-stick is below a 2/2-stick the children had made.)
- T: Take a piece that will be the same size as the 1/2, but it will be a different fraction. You remember that last time we had 3/15 that was also 1/5?
- L: (Moves the copy she made around with the cursor.)
- J: I think I know how.
- T: (To Laura) look and see what he is doing.
- J: (Dials PARTS to 4 and clicks on the copy of the unit stick beneath the 1/2-stick. He then pulls two parts out of this 4/4-stick, making a 2/4-stick.)
- T: That's beautiful!! How would you label it?
- J: 2/4.
- T: So, now we have 1/2 and 2/4. Another member of the family, please?
- L: (Grabs the mouse and makes a copy of the unit stick.) Do we only make 1/2?
- T: I didn't hear you.
- J: I know four of them.
- L: (Dials PARTS to 5) do we have to put 1/2 right now (that is, begin with 1/2)?
- T: The family we are after is the 1/2 family (an evasive answer).
- J: I know more than 6!
- L: Can I put 10 on here (the dial of PARTS)?
- T: Go ahead, as long as you can take a piece that is like 1/2, that's Ok.
- L: (Dials PARTS to 10, clicks on a copy of the unit stick, drags the 2/4-stick Jason made to the vicinity of the 10/10-stick and erases the mark on it.

She then drags the resulting 1/2-stick directly over the 10/10-stick with left endpoints aligned. She then looks at the teacher as if done.)

T: Can you pull out the part that you think is one half?

L: Yes. (Using the 1/2-stick as a guide, she pulls a 5/10-stick from the 10/10-stick.)

T: So, how much is it of the big cake?

L: 5/10 (She then labels the stick "5/10" using LABEL).

J: (As Laura is labeling the 5/10-stick) I know a lot of them. There are more than 6! I can do it for 100, 50, ...

T: (To Laura) How did you find 5/10?

L: Because I know that 1/2 of 10 is 5.

Jason's comment, "I know a lot of them. There are more than 6! I can do it for 100, 50, ..." solidly indicates that he dropped out the necessity of actually carrying out recursive partitioning. He posited possible partitionings and focused on the numerosity of the parts of those partitionings. Saying "I can do it for 100, 50, ..." solidly indicates that it was no longer necessary for him to actually carry out recursive partitioning in the situation. Initially, however, Laura didn't know what to do—"What am I supposed to do?"—and generated 5/10 only after Jason made 2/4. After Jason made 2/4, she seemed to establish a goal but it certainly cannot be said to involve recursive partitioning. Rather, her goal seemed to be related to Jason's partitioning of a copy of the unit stick into four parts. Her question, "Do we have to put 1/2 right now?" and her subsequent act of partitioning the stick into 10 parts together indicate that her goal wasn't to recursively partition the 2/2-stick. Rather, it was to find a number (10) that she could partition into two equal parts ("I know that 1/2 of 10 is 5!"). Laura abstracted no general way of operating, as did Jason. Rather, she selected a number of which she could find 1/2.

Following Laura's lead, in the next task of the teaching episode, Jason made an 18/18-stick and then pulled nine parts of the stick out to make a fraction commensurate with 1/2. He gave the same type of explanation for why he chose 18 that Laura gave for why she chose 10. Laura then, quite strongly, asserted that she knew one and chose 48. However, she didn't know what 1/2 of 48 was and resorted to counting from 1 to segment 48 into two equal parts. After a long pause Jason asserted he got it, but the teacher never asked him what he got or how he arrived at it. On the teacher's suggestion, Laura dragged a 1/2-stick directly over the 48/48-stick and used it as a guide to pull out 24 parts. She then said that was 24/48 and used LABEL to label it "24/48." While she was completing her labeling activity, Jason said, "I know a big one!"

Protocol VII: Jason Using Recursive Partitioning to Make a 200/200-Stick.

J: I know a big one! (Copies a unit stick and activates PARTS) How many does this one go to?

T: Let's see if you remember.

J: (Groans.)

T: How many do you want?

J: 200!

T: Well, think of a way to make 200!

J: (Dials PARTS to 50 and clicks on the stick. He then dials PARTS to 4 and clicks on each one of the 50 parts of the 50/50-stick.)

T: (To Laura) what do you think he is going to do (While Jason is dialing 4)?

L: I don't know. ...

T: (To Laura) How many parts do you think he is going to have (after Jason dialed PARTS to 4)? Is it going to be 200?

L: Yeah, because 50 times 4 is 200.

T: Can you think of a way to make 200 that will not take you so long to do it?

L: (After a long pause) I know a way to make 100. That will be 50 and 50 and 50 and 50.

T: How much did you say?

L: 200.

T: Well, then try it.

J: That's what I did.

T: Yeah. But can you think of a way to make 200 instead of making all these 4s again, again, and again?

J: I know how (makes a copy of the unit stick, dials PARTS to 4 and clicks on the stick. He then dials PARTS to 50 and clicks on the first two parts, but is interrupted by the teacher).

T: Stop here, because you have the half and if you are going to do that whole thing, you are going to have to click how many times?

J: 4!

T: After you finish, you will have 200 pieces. How many pieces would you take out in order to have a half of 200?

J: 100.

T: How many pieces will you have to do (pull out)?

J: 100.

T: Are you going to do that?

J: (Shakes his head "no.")

T: Can you think of a way to take 100 pieces?

J: (Cuts the stick at the midpoint.)

T: That's good. (Asks Jason to label it. Jason creates the label "100/200" using LABEL.)

T: (To Laura) Which one are you going to make?

L: (Dials PARTS to 3 and clicks on a copy of the stick in the RULER. She then dials PARTS to 5 and clicks on each of the three parts.) Half of that will be (aligns the 1/2-stick directly beneath the 15/15-stick with left endpoints coinciding). Oh, it doesn't have a half!

T: (To Jason) You said earlier that you know of 6, and then you said that there are a lot more than 6. How many do you think there?

J: There's like 600, something like that.

T: Why do you think there are so many?

J: There are many numbers that you can put. ... No there are more than 600! 100,000 or something like that.

Jason's choice of 200 and his partitioning the stick into 50 parts and then each of the 50 parts into four parts after he realized that he couldn't dial PARTS to 200 indicates the power recursive partitioning held for him. Had he first partitioned the stick into four parts and then each part into 50 parts, it wouldn't have been as dramatic as the other way around because that could be based on knowing that $50 + 50$ is 100, so $50 + 50 + 50 + 50 = 200$. In fact, Laura made 200 in just this way, but there was yet no indication that she engaged in recursive partitioning. The power of his abstractions concerning making 1/2 of the unit stick is illustrated by his comments in the last part of Protocol VII after the teacher asked him how many (fractions) he thought there were. He even thought there were more than 600—"100,000 or something like that."

For Jason, 100,000 was a sort of unbounded number; it is greater than any number with which he could operate in a meaningful manner. Thus, his recognition that there are 100,000 fractions commensurate with 1/2 indicates that he had established the existence of a rather unbounded plurality of such fractions. As noted in the discussion concerning Protocol VI, Jason no longer needed to carry out the operations of recursive partitioning to produce fractions commensurate with 1/2, but was now operating symbolically, taking for granted the results of recursive partitioning. This is a crucial step for him to establish a class of fractions commensurate to 1/2. Although it is problematic whether Jason actually constructed this class, I distinguish the plurality of fractions commensurate with 1/2 from the production of just a few such fractions. Jason's plurality was an abstraction from past and future (anticipated) productions, and production itself was no longer needed—only the possibility of production. We should note that Jason's plurality, though unbounded, is probably not exhaustive. His claim that "there are many numbers that you can put" indicates the plurality but also stands in contrast to the absence of a claim that he could put any number.

Some indications emerged that Laura began to engage in recursive partitioning. After the teacher asked her which one she was going to make, she partitioned the stick into three parts and then each part into five parts. Although her partitioning activity was carried out somewhat independently, Jason's partitioning the stick into 50 parts and then each of these parts into four parts could be assimilated using her units-coordinating scheme. I would say that her partitioning activity under discussion was produced as a result of a generalizing assimilation of her units-coordinating scheme as she experienced Jason's partitioning activity, which excludes imitation as an explanation of her repartitioning activity. It wasn't simply an imitation because she had to decide to partition the stick into three parts, and then each one of the three parts into five parts. This was not a given in Jason's activity.

DISCUSSION

This account of the construction of commensurate fractions spans the teaching episodes held on October 25 and November 8, 15, and 22 of the children's fifth grade in school. Because the teaching experiment continued on throughout the children's fifth grade, it is by no means a complete account of the two children's construction of commensurate fractions (cf. Steffe, 2003, for an account of the continuation of the teaching experiment with Jason and Laura). However, I have tried to say enough so that my stance that learning trajectories of children must be constructed by teacher/researchers who participate first-hand in children's constructive activity can be taken seriously. These learning trajectories have far reaching consequences for the mathematical education of children because regarding a fraction as being produced as a result of using a scheme of a certain kind is quite different than considering fractions in a context of "fractional models," such as the set model, the region model, fraction strips, or the number line model. These models are often used by teachers in attempts to convey meaning for fractions and fractional operations to the children they teach.

On the Origins of Children's Fractional Knowledge

In contrast, Jason's learning trajectory demonstrates that children, as a result of their productive thinking, can construct fractions, fractional multiplication, and commensurate fractions. This demonstration opens the possibility of turning the issue of the source of meaning of fractions and their operations from being conveyed to children using fractional models to being supplied by children as the result of using a particular kind of fractional scheme as they engage in interactive mathematical communication. It is critical to note that the source of Jason's productive thinking resided in his use of his numerical concepts and operations in establishing fractional concepts and operations. In fact, the major operations of the partitive fractional scheme—partitioning, iterating, and disembedding—have their genesis in the child's numerical concepts. For example, marking off one of four fair shares of a stick is accomplished by mentally projecting four units into the stick and testing if a selected part is a fair share is accomplished by iterating the part to generate a stick that can be compared to the original stick. These operations were established by Jason using his composite unit, 4, as a template for partitioning and the iterability of the units of 1 it contains as an operation to iterate the marked share. So, in any learning trajectory for commensurate fractions, a very important starting point is for children to have constructed the units of their numerical concepts as iterable units so they do not only use their numerical concepts in making one of so many fair shares of unit wholes, but also so that they test whether the share is fair by iterating the share.[11]

At this level in children's construction of numerical concepts, children can conceive of partitioning a stick into a rather large number of parts and so can conceive of fractions like 1/250 or 100/250 as did Jason when he used recursive partitioning to partition a stick into 200 parts. Because PARTS could be dialed to 99, the children could use PARTS to instantiate their numerical concepts up to 99, and beyond in the case recursive partitioning was available as templates to partition a stick. It was a distinct advantage that TIMA: Sticks was programmed in such a way that the computer actions could be used by the children in instantiating their conceptual operations because the children were not limited to partitioning a stick into only a rather small number of parts nor were they limited to only imagining pulling, say, three parts out of five equal parts of a pizza. In the case of actually taking three of five equal parts of a pizza, the pizza has to be disassembled. In TIMA: Sticks, if a stick is regarded as a representative of a pizza, not only can the child precisely partitioning the stick into five parts, the child can also pull out the same part three times or three parts once while leaving the stick intact. So, the conceptual operation of disembedding can be instantiated by the child as well as iterating a part along with other operations that are relevant, such as breaking apart, copying parts, or joining parts. It was the case that in TIMA: Sticks, only segments served as material of operating. But, we were more interested in engendering the children's ways and means of operating in their construction of fractional schemes than in partitioning different kinds of unit wholes. Although Piaget, Inhelder, & Szeminska (1960) did find that children's ability to partition areas into thirds constitutes a very significant developmental advance, their findings were not replicated with Jason and Laura primarily because of these children's rather advanced numerical concepts.[12] Moreover, it should be said that Piaget et al. were more interested in the conceptual operations that were developing than they were in the material of the operations.

Two Learning Levels for Fractional Schemes

The distinction that I have made between the equi-partitioning scheme and the splitting operation essentially defines two distinct learning levels for fractions. The equi-partitioning scheme was observed at the end of Jason's third grade in school,

---

[11]See Steffe (2002) for an elaboration the iterative operation as it pertains to units of 1.

[12]Although I do not give an account of it, we used TIMA: Bars with the children in which the child makes and operates with rectangular regions. Using this computer tool, the children could partition the rectangular regions horizontally and vertically. There were no obvious distinctions in the children making one-dimensional partitions in TIMA: Bars and making partitions of sticks in TIMA: Sticks.

and he spent his entire fourth grade transforming that scheme into his partitive fractional scheme. Various aspects of his partitive fractional scheme have been documented (Steffe, 2002; Tzur, 1999). For example, if Jason produced 2/5 of a stick, he knew that the fractional part of the stick left was 3/5 of the stick. However, there were internal constraints within his partitive fractional scheme that were inherited from his equi-partitioning scheme, and these constraints were manifest as his inability to produce, say, 13/7 of a stick, a fraction commensurate with a given fraction, and even the most elementary split of a stick—a two-split (Steffe, 2002). So, it was a major surprise that Jason independently produced 3/16 as the fractional part of a stick that 3/4 of 1/4 of the stick made in October 25 teaching episode. To explain Jason's advancement, I introduce the concept of recursive partitioning. But the way that I explain this concept does not emphasize that Jason used his whole number multiplying scheme in his initial production of a fractional multiplying scheme that I refer to as fractional composition.

In the continuation of Protocol IV, Jason insightfully partitioned a stick first into three parts and then each part into four parts to produce a 12-part stick. I refer to his partitioning activity as recursive partitioning, but it can be also interpreted as a coordination of the units of 3 and 4. In the case of discrete units, a coordination of these two units can be thought of as placing four units of some kind into each of three units, such as placing four candies into each of three bags. In the partitioning context, to make a 12-part stick, Jason first partitioned a stick into three parts and then inserted partitioning into four parts into each of these three parts. This constituted a generalizing assimilation of units-coordinating that was manifest in Protocol III when Jason explained how he produced 3/16 given 3/4 of 1/4: "See, if we would have had it in that ... 4,4,4, and 4—16." Jason definitely embedded using his units-coordinating scheme (his whole number multiplying scheme) in what was by this time a reversible partitive fractional scheme to produce what might conventionally be referred to as fractional multiplication. However, I referred to Jason's productive activity as a fractional composition scheme because Jason had no sense of engaging in multiplicative activity. From his perspective, he was solving a problem involving a fraction of a fraction. Moreover, when Jason used his whole number multiplying scheme in both of Protocols III and IV, the indication is solid that he did not regard what he was doing as multiplication, especially in Protocol III when he found that 3/16 was 3/4 of 1/4. He knew well that $4 \times 4 = 16$ without reasoning. That he did engage in reasoning to produce "16" solidly indicates that although he used units-coordinating operations to achieve his nonpartitioning goal, his use of the operations was novel and constituted a creative act. His novel and creative use of his units-coordinating operations in the service of a nonpartitioning goal is the primary reason that I referred to his use of them as recursive partitioning.

Laura demonstrated in both of Protocols III and IV that even though she had constructed a units-coordinating scheme and had used it extensively in her fourth grade to partition the elements of a prior partitioning, the availability of the scheme is insufficient for it to be evoked by the goal of finding how much 3/4 of 1/4 of a stick is of the stick or by the goal of partitioning a stick into 12 parts without explicitly using 12 in the partitioning. What this means is that it is the child who introduces the use of units coordinating into the construction of a fractional composition (or multiplying) scheme as a creative and novel act. The splitting operation seems to be the essential enabler of this creative and novel act.

At the outset of the teaching experiment, I predicted that the splitting operation would be necessary in the construction of commensurate fractions. Without mentioning the splitting operation, however, I used recursive partitioning as the explanatory construct in Jason's production of the fractional compositional scheme in October 25 teaching episode.

Jason used recursive partitioning throughout the four teaching episodes that were held with the children during the month of November, and it was this operation that served as the basic source of his productive and independent mathematical activity. Although I did not mention the splitting operation in my explanation of recursive partitioning, I consider the former as implicit in the latter. When it was Jason's goal to find how much three small parts (3/4 of 1/4) of a stick were of the whole stick, I assumed that he regenerated an image of the 4/4-stick and considered the three small parts as possible parts of it. Because his units of 1 were iterable, it would be possible for him to conceive of the whole stick as composed of an indefinite numerosity of small parts without actually engaging in iterative activity. He was aware that he could engage in such iterative activity, so he didn't need to run through the activity to realize its potential results, which means that the potential results were available prior to operating. That is to say, he anticipated splitting the whole stick into an indefinite numerosity of small parts.

Anticipating splitting the whole stick into an indefinite numerosity of small parts could be also realized if Jason anticipated iterating the partitioned 1/4 of the stick four times to partition the whole of the 4/4-stick. These operations also constitute splitting operations with the important proviso that a composite unit of 4 is involved as well as the unit of 1. In either case, the splitting operation is implicit in the fractional composition scheme, and it is a constitutive aspect of recursive partitioning. In fact, it was the splitting operation that allowed me to consider Jason's fractional composition scheme as a metamorphosis of his partitive fractional scheme. My claim that the fractional composition scheme is a level of abstraction above the partitive fractional scheme is corroborated by Jason's construction of the commensurate fractional scheme in the four teaching episodes analyzed in this article along with his inability to construct this scheme while he was in the fourth grade (Steffe, 2002). The claim is also compatible with the observed differences between the two children in the teaching episodes. So, what we have in the cases of Jason and Laura are two children who were operating at distinctly different learning levels with respect to the splitting operation.

Laura's Lack of Producing Independent Explanations

Laura did not construct recursive partitioning over the duration of the month of November in spite of the best attempts of the teacher to bring it forth. Apparently, the two operations of partitioning and iterating that were available to Jason opened possibilities for him that were not present for Laura. The difference in the two children was striking. For example, in the teaching episode held on November 8 (Protocol IV), to partition a stick into 12 parts without using 12, Laura partitioned the stick into 11 parts and then pulled one part from the stick and joined it to the 11-part stick. Jason, on the other hand, first partitioned the stick into three parts and then each of these three parts into four parts. In this case, Laura did engage in independent mathematical activity, but her way of operating constituted contraindication of recursive partitioning operations. It can be conjectured that Laura simply didn't think to operate in the way Jason operated because once her units-coordinating scheme was activated in the situation by means of observing Jason operate, she did know what to do and proceeded quite smoothly. However, it was characteristic for Laura to need to reenact an explanation or a recursive partitioning made by Jason, or for there to be visual cues in her perceptual field before she could engage in the actions that were needed to be successful in explaining why a fraction such as 1/3 was commensurate to, say, 4/12 after she measured the 4/12-stick (Protocol IV: Continued). Jason could independently engage in the operations that were necessary to produce such explanations or actions, and beyond that, he could independently produce a unit fraction that was commensurate with, say, 3/15 (Protocol V). I consider such independent productions as necessary to judge that a child has constructed a scheme for producing a fraction commensurate with a given fraction.

Throughout the first four teaching episodes, the teacher attempted to bring forth recursive partitioning within Laura. However, as I have just indicated, Laura could use her units-coordinating scheme in recognition or in the reenactment of prior recursive partitioning actions or explanations made by Jason. But she could not use it in the absence of such elements. For example, in Protocol V, after Laura had made a 15/15-stick by first partitioning a unit stick into three parts then each part into five parts, the teacher pulled out three parts of the 15/15-stick in a test to find if Laura could explain why 3/15 is also 1/5 after the teacher said that it was 1/5. Laura said that she didn't agree that it could be 1/5, and it wasn't until Jason explained that it took five of the 3/15-sticks to make the 15/15-stick by pulling a 3/15-stick from the 15/15-stick and making copies and aligning five 3/15-sticks directly beneath the 15/15-stick that she said, "I get it!" Her saying that she got it was made possible by her units-coordinating scheme and that she knew that $3 \times 5 = 15$. However, using her units-coordinating scheme to establish a post hoc understanding of Jason's explanation is quite different than using the scheme in a creative production of an explanation. This difference is especially apparent when Laura, in the very next task, looked disconcerted when 1/5 appeared in the number box upon measuring a 3/15-stick she had pulled out from a 15/15-stick.

Given the constraint that Laura could not independently produce an explanation for why 3/15 was also 1/5, the teacher changed the task situation to asking the children to produce fractional sticks commensurate to a 1/2-stick (Protocols VI and VII). The expectation of the teacher was that Laura would use her units-coordinating scheme to produce a sequence of fraction sticks each commensurate with the 1/2-stick. Similar to other situations, Laura initially asked, "What am I supposed to do?" Jason, on the other hand, said, "I think I know how" and proceeded to make a 2/4-stick by using recursive partitioning. The experiment failed in the case of Laura in that she partitioned the unit stick into 10 parts and took out 5 because she knew that 5 is 1/2 of 10. This way of operating, however, did constitute productive activity on Laura's part because she then posited 48 as a possibility. So, when Laura independently generated a situation that was made possible by her part-whole fractional scheme, she became playfully involved in a way that was quite similar to the way Jason became playfully involved. For example, when Jason made the 200/200-stick by partitioning a unit stick into 50 parts and then each part into four parts, one could legitimately say that he engaged in independent mathematical activity with a playful orientation, which I regard as mathematical play (Steffe & Weigel, 1994). Laura could and did engage in mathematical play, but her possibilities were more restricted than were Jason's.

Vertical Learning and Lateral Learning

It was the case that Laura did not provide any indication of the splitting operation, recursive partitioning, or a fractional composition scheme over the duration of the teaching episodes. However, she did engage in mathematical interaction with Jason in those cases where her units-coordination scheme was called forth in recognition of Jason's language and actions. But this was not enough for her to engage in making independent explanations for why a fraction was commensurate to another fraction, and she remained dependent on what Jason said or did, what the teacher said or did, or the context of the situation to know what to do to be successful. The issue this raises for mathematics teaching is a fundamental issue because it is not uncommon for students to be in a similar position with respect to other situations of learning. Should a teacher continue to engage Laura in tasks that she cannot independently solve, or switch to tasks that Laura can solve independently? The question should not be interpreted as engaging Laura in tasks that could engender learning versus tasks that she could solve without learning being involved. Although Jason could independently produce fractions commensurate with a given fraction, he did not know that, say, 1/5 was another fractional name for 3/15 prior to experience. Nor could he engage in reasoning to produce, say, 1/3 as another fractional name for 4/12 without enacting his operations in a context. So, Jason needed to engage in what I call lateral learning, that is, learning that involves using his current schemes in novel ways to establish new schemes whose operations can be judged to be more or less at the same level as the operations of his current schemes (e.g., the partitive fractional scheme was established as a modification of his equi-partitioning scheme), or learning that involves establishing specific results of using a current scheme through reflection and abstraction. He also needed to engage in what I call vertical learning so that he could mentally engage in the operations that produce a class of fractions equivalent to a given fraction without operating on sensory material in a way that is analogous to how he produced a plurality of fractions commensurate to 1/2 in Protocol VII.

Analogously, Laura was in the position of needing to engage in vertical learning in the construction of the splitting operation, so the question of whether to continue on engaging her in commensurate fractional tasks is equivalent to asking how a teacher can engender vertical learning. This question can be answered only in retrospect, that is, after vertical learning has been observed of the kind that enabled Jason to produce his fractional composition scheme as a creative act. But there is no history of Jason's lateral learning that would contribute to understanding his construction of his fractional composition scheme that is available to me to which I can refer to develop a sequence of learning tasks that, if solved by a child who has constructed the reversible partitive fractional scheme, would likely bring forth the splitting operation within the child. I presently consider the splitting operation to be the result of a developmental metamorphic accommodation of the reversible partitive fractional scheme.[13] A developmental metamorphic accommodation occurs as a global result of using a scheme and occurs independently. It reconstitutes the scheme on a new level and reorganizes the scheme at that level and so it is a result of reflective abstraction (Steffe, 1994).

ACKNOWLEDGMENTS

The National Science Foundation funded the research on which the article is based as part of the activities of NSF Projects RED–8954678 and REC–9814853. All opinions are those of the author.

This article was prepared for the symposium, *The Use of Learning Trajectories in Curriculum Development and Research*, held at the Research Presession to the

---

[13]I remind the reader that the partitive fractional scheme is an elaboration of the equi-partitioning scheme and so contains the operations of partitioning and iterating.

2002 Annual Meeting of the National Council of Teachers of Mathematics, Las Vegas, Nevada.

REFERENCES

Olive, J., & Steffe, L. P. (2002). The construction of an iterative fractional scheme: The case of Joe. *Journal of Mathematical Behavior, 20,* 413–437.

Piaget, J., Inhelder, B., & Szeminska, A. (1960). *The child's conception of geometry.* New York: Basic Books.

Simon, M. (1995a). Elaborating models of mathematics teaching: A response to Steffe and D'Ambrosio. *Journal for Research in Mathematics Education, 26,* 160–162.

Simon, M. (1995b). Reconstructing mathematics pedagogy from a constructivist perspective. *Journal for Research in Mathematics Education, 26,* 114–145.

Steffe, L. P. (2002). A new hypothesis concerning children's fractional knowledge. *Journal of Mathematical Behavior, 102,* 1–41.

Steffe, L. P. (2003). Fractional commensurate, composition, and adding schemes: Learning trajectories of Jason and Laura: Grade 5. *Journal of Mathematical Behavior, 22,* 237–295.

Steffe, L. P., & D'Ambrosio, B. (1995). Toward a working model of constructivist teaching: A reaction to Simon. *Journal for Research in Mathematics Education, 26,* 146–159.

Steffe, L. P., & Olive, J. (1990). *Children's construction of the rational numbers of arithmetic.* Proposal to the National Science Foundation. The University of Georgia, Athens: Author.

Steffe, L. P., & Thomson, P. (2000). Teaching experiment methodology: Underlying principles and essential elements. In A. E. Kelly & R. A. Lesh (Eds.), *Handbook of research design in mathematics and science education* (pp. 267–303). Hillsdale, NJ: Lawrence Erlbaum Associates, Inc.

Steffe, L. P., & Wiegel, H. G. (1994). Cognitive play and mathematical learning in computer microworlds. *Journal of Research in Childhood Education, 8,* 117–131.

Steier, F. (1995). From universing to conversing: An ecological constructionist approach to learning and multiple description. In L. P. Steffe & J. Gale (Eds.), *Constructivism in education* (pp. 67–84). Hillsdale, NJ: Lawrence Erlbaum Associates, Inc.

Tzur, R. (1999). An integrated study of children's construction of improper fractions and the teacher's role in promoting that learning. *Journal for Research in Mathematics Education, 30,* 390–416.

MATHEMATICAL THINKING AND LEARNING, 6(2), 163–184

Young Children's Composition of Geometric Figures: A Learning Trajectory

Douglas H. Clements
Department of Learning & Instruction
University at Buffalo, State University of New York

David C. Wilson
Department of Mathematics
University at Buffalo, State University of New York

Julie Sarama
Department of Learning & Instruction
University at Buffalo, State University of New York

The purpose of this research is to chart the mathematical actions-on-objects young children use to compose geometric shapes. The ultimate goal is the creation of a hypothetical learning trajectory based on previous research, as well as instrumentation to assess levels of learning along the developmental progression underlying the trajectory. We tested both the developmental progression and the instrument through a series of studies, including formative studies (including action research by 8 teachers) and a summative study involving 72 children ages 3 to 7 years. Results provide strong support for the validity of the developmental progression's levels and suggest that children move through these levels of thinking in developing the ability to compose 2-dimensional figures. From lack of competence in composing geometric shapes, they gain abilities to combine shapes—initially through trial and error and gradually by attributes—into pictures, and finally synthesize combinations of shapes into new shapes (composite shapes).

Requests for reprints should be sent to Douglas H. Clements, Department of Learning & Instruction, University at Buffalo, State University of New York, Graduate School of Education, 505 Baldy Hall (North Campus), Buffalo, NY 14260. E-mail: clements@buffalo.edu

Up to the eighth or ninth year of childhood one may say that the child does hardly anything else than handle objects, explore things with his hands, doing and undoing, setting up and knocking down, putting together and pulling apart; for, from the psychological point of view, construction and destruction are two names for the same manual activity. Both signify the production of change, and the working of effects, in outward things. The result of all this is that intimate familiarity with the physical environment, that acquaintance with the properties of materials things, which is really the foundation of human *consciousness*. (James, 1958, pp. 53–54)

The ability to describe, use, and visualize the effects of composing and decomposing geometric shapes is a major conceptual field and set of competencies in the domain of geometry. Although a few studies address children's thinking about geometric composition, there is little or no research detailing developmental progressions (Simon, 1995, p. 133), much less complete hypothetical learning trajectories (creation of which is the ultimate goal of our work; see our introduction to this special issue). Here we emphasize our theory of developmental progression, which posits that children move through several distinct levels of thinking and competence in the domain of composition and decomposition of geometric figures. These competencies include the child's ability to make pictures or designs with shapes (moving from trial and error to use of attributes), create and maintain a shape as a unit, and combine this shape with another shape to create a new shape that is conceptualized as an independent entity. We concurrently developed an assessment instrument to measure the levels of thinking evinced by prekindergarten through second-grade children. In this article, we describe our initial development and formative research and fully report a large study evaluating the validity of the developmental progression in shape composition, determining whether the hypothesized levels do indeed form a disjunctive scale and evaluating the validity and reliability of the assessment instrument. We briefly describe the context and the development of all components of a hypothetical learning trajectory.

BACKGROUND

The context for the work is a software and print curriculum development project, *Building Blocks—Foundations for Mathematical Thinking, Pre-Kindergarten to Grade 2: Research-based Materials Development*, funded by the National Science Foundation. Our intention is to make mathematical software that is motivating and educationally effective. This is a multiyear project involving nine phases of research and development (Clements, 2002; Clements & Battista, 2000; Sarama, 2004); here, we summarize only the most relevant phases. First, the existing research on mathematics teaching and learning for prekindergarten through second grade is reviewed and synthesized so that it can be directly applied to the development of the software and print curricula. A tentative developmental progression is then created for each major strand of the project, including the domains of number (i.e., counting, comparison, subitizing, disembedding/composition of units, addition/subtraction, and multiplication/division) and geometry (i.e., shapes, composition of units, imagery/transformations/congruence, symmetry, coordinates, and measurement). Next, activities are written that are hypothesized to guide children through each of the levels of this developmental progression, both as a group and as individuals. The initial drafts are revised based on input from various project participants including classroom teachers. The activities are then pilot- and field-tested with groups of increasing size (individual to classroom), with revisions during and after each test. The result is a hypothetical learning trajectory, including "the learning goal, the learning activities, and the thinking and learning in which the students might engage" (Simon, 1995, p. 133). In this way, learning trajectories (e.g., see Clements, 2002; Cobb & McClain, in press; Gravemeijer, 1999; Simon, 1995) stand at the heart of our research and development model.

The mathematics goal of the learning trajectory discussed here is the ability to describe, use, and visualize the effects of composing and decomposing geometric shapes. This domain is significant in that the concepts and actions of creating and then iterating units and higher order units in the context of constructing patterns, measuring, and computing are established bases for mathematical understanding and analysis (Clements, Battista, Sarama, & Swaminathan, 1997; Reynolds & Wheatley, 1996; Steffe & Cobb, 1988). Additionally, there is empirical support that this type of composition corresponds with, and supports, children's ability to compose and decompose numbers (Clements et al., 1996). In contrast to other approaches (Gravemeijer, 1994), after establishing the goal, we developed a first version of a developmental progression for the domain (Clements, 2002; Sarama & Clements, in press).

DEVELOPMENT AND FORMATIVE RESEARCH

Initial Observations

The genesis of the shape composition learning trajectory was in observations made of children using *Shapes* software (Sarama, Clements, & Vukelic, 1996) to compose shapes. Shapes (Clements & Sarama, 1998) is a computer manipulative, a software version of pattern blocks, that extends what children can do with these shapes. Children create as many copies of each shape as they want and use computer tools to move, combine (compose and decompose), and duplicate these shapes to make pictures and designs and to solve problems, such as completing puzzles (i.e., filling outlines with shapes). Sarama et al. (1996) observed that several case-study children followed a similar progression in choosing and combining shapes (e.g., rhombi or equilateral triangles) to make another shape (e.g., to fill a hexagonal frame). Initially, children merely appreciated the relation between pattern blocks (e.g., how one pattern block could be made using other pattern blocks), but their efforts to fill a hexagonal frame with other pattern blocks was by trial-and-error. Later, they recognized the hexagon could be made with two trapezoids, followed by other combinations. Sarama et al. reviewed the behaviors all kindergarten children exhibited when composing shapes and found several similar sequences and noted that, throughout the study, children's development appeared to move from placing shapes separately to considering shapes in combination— from manipulation- and perception-bound strategies to the formation of mental images (e.g., decomposing shapes imagistically); from trial and error to intentional and deliberate action and eventually to the prediction of succeeding placements of shapes; and from consideration of visual "wholes" to a consideration of side length, and, eventually, angles.

Through cycles of curriculum revision and observations, we created the first draft of the hypothetical learning trajectory, including the developmental progression Sarama et al. (1996) observed and a set of tasks (including on- and off-computer puzzles) that appeared to facilitate growth for children at different points along the trajectory (as embodied now in the Shapes software and the corresponding print curricula). Following this curriculum development effort, we combined these observations with related observations from other researchers (Mansfield & Scott, 1990; Sales, 1994) and some elements of psychological research (e.g., Vurpillot, 1976) to refine this hypothetical learning trajectory. This article focuses on the refinement of the developmental progression.

Developmental Progression

Several theoretical assumptions underlie the refined developmental progression. First, to solve certain manipulative shape composition tasks effectively and efficiently, children must build an image of a shape and then match that image to the goal shape by superposition (of both components and shapes), performing mental transformations as necessary to match these images. Second, children's knowledge of shapes develops from little knowledge, to syncretic knowledge (a global combination of perceptions without analysis), to the conscious ability to recognize, describe, and manipulate not only individual shapes, but also their components, and eventually, their properties (Clements, Battista, & Sarama, 2001). This last development is related to their developing knowledge of shapes which Pierre and Dina van Hiele (van Hiele, 1986) have theorized develops from a generic visual impression to property recognition and hierarchical classifications. The developmental progression goes beyond existing van Hielian thought in adding the composition and decomposition processes as essential elements of geometric knowledge. These processes include the child's ability to create and maintain a shape as a unit, combine this shape with another shape (initially by trial and error, then by considering attributes), then apply a uniting operation to reconceptualize the composite shape as a new unit. These processes operate first on physical shapes and later on mental constructs. Consistent with our interpretation of the van Hiele theory, children's ability to combine shapes develops from a trial-and-error combination of whole shapes through an increasing ability to combine shapes based on their attributes (components, such as side lengths, and, later, properties, such as angle size).

Table 1 describes the seven levels that constitute the developmental progression for the composition of shapes. Note that to save space, we provide only our most recent version of these descriptions.

Instrument

Based on this developmental progression, an instrument was designed to assess each of the levels of thinking. The items contained in the instrument were adapted from a wide variety of sources (e.g., state and national tests, curricula) or constructed by researchers and were agreed on by several colleagues, project investigators on one of the funded projects of which this study was a part.[1] The following are selected examples of 3 of the 19 items on the instrument accompanied by descriptions of how the children's responses were assessed in terms of the developmental progression.

1.   Children were given pattern blocks and a frame of a "man" (Fig. 1a) and asked to "use pattern blocks to fill this puzzle." Children were categorized as follows: Precomposer: cannot match shapes to well-defined, frames, such as the "feet;" Piece Assembler: can fill well-defined frames (e.g., the "feet" only) using trial and error; Picture Maker: partially fills frames by trial-and-error, matching shapes by boundary or matching side lengths; Shape Composer: completes entire frame with deliberate choices of shapes by matching configurations, sides, or angles; Substitution Composer: deliberately replaces a group of shapes (e.g., two triangles) with one shape (e.g., blue rhombus) or vice versa; and Shape Composite Iterater: deliberately, systematically iterates a composite group of shapes to fill a region.

2.   Children were asked, "If these shape pieces [examiner gestured to the top four shapes in Fig. 1b] were pushed together so their sides touch, which big shape [examiner gestured to the bottom four figures] would they make?" If correct on this item (adapted from Lane & Rosser, 1987), children were considered to be operating on the Shape Composer level.

[1]This included the following researchers from the National Science Foundation grant, "Technology-enhanced Learning of Geometry in Elementary Schools:" Daniel L. Watt, Richard Lehrer, Richard Lesh, and Zuzka Blasi.

TABLE 1
Levels in the Developmental Progression for the Composition of Shapes

---

Precomposer. Children manipulate shapes as individuals, but are unable to combine them to compose a larger shape. For example, children might use a single shape for a sun, a separate shape for a tree, and another separate shape for a person. Children cannot accurately match shapes to simple frames (closed figures that can be filled with a single shape).

Piece Assembler. Children at this level are similar to Precomposers, but they place shapes contiguously to form pictures. In free form "make a picture" tasks, for example, each shape used represents a unique role, or function in the picture (e.g., one shape for one leg). Children can fill simple frames using trial and error (Mansfield & Scott, 1990; Sales, 1994), but have limited ability to use turns or flips to do so; they cannot use motions to see shapes from different perspectives (Sarama et al., 1996). Thus, children at the first two levels view shapes only as wholes and see few geometric relationships between shapes or between parts of shapes (i.e., a property of the shape).

Picture Maker. Children can concatenate shapes contiguously to form pictures in which several shapes play a single role (e.g., a leg might be created from three contiguous squares), but use trial and error and do not anticipate creation of new geometric shapes. Shapes are chosen using gestalt configuration or one component such as side length (Sarama et al., 1996). If several sides of the existing arrangement form a partial boundary of a shape (instantiating a schema for it), the child can find and place that shape. If such cues are not present, the child matches by a side length. The child may attempt to match corners, but does not possess angle as a quantitative entity, so they try to match shapes into corners of existing arrangements in which their angles do not fit (a "picking and discarding" strategy). Rotating and flipping are used, usually by trial-and-error, to try different arrangements. Thus, they can complete a frame that suggests that placement of the individual shapes but in which several shapes together may play a single semantic role in the picture.

Shape Composer. Children combine shapes to make new shapes or fill puzzles, with growing intentionality and anticipation ("I know what will fit"). Shapes are chosen using angles as well as side lengths. Eventually, the child considers several alternative shapes with angles equal to the existing arrangement. Rotation and flipping are used intentionally (and mentally, i.e., with anticipation) to select and place shapes (Sarama et al., 1996). They can fill complex frames (figures whose filling requires multiple shapes; Sales, 1994) or cover regions (Mansfield & Scott, 1990). Imagery and systematicity grow within this and the following levels. In summary, there is intentionality and anticipation, based on the shapes' attributes, and thus, the child has imagery of the component shapes, although imagery of the composite shape develops within this level (and throughout the following levels).

Substitution Composer. Children deliberately form composite units of shapes (Clements et al., 1997) and recognize and use substitution relations among these shapes (e.g., two trapezoid pattern blocks can make a hexagon).

Shape Composite Iterator. Children construct and operate on composite units (units of units) intentionally. They can continue a pattern of shapes that leads to a "good covering," but without coordinating units of units (Clements et al., 1997).

Shape Composer with Superordinate Units. Children build and apply (iterate and otherwise operate on) units of units of units.

---

3.   Children were asked to "use pattern blocks to fill these puzzles [Fig. 1c]. Fill each puzzle a different way." Children were categorized into the Precomposer to Shape Composer levels in ways similar to the first example, but this item was designed especially to target the level of Substitution Composer. To be so categorized, children would have to deliberately replace certain blocks with others from one design to the next.

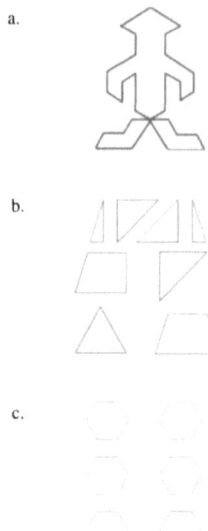

FIGURE 1 Example items from the instrument.

Because we did not wish to reify the developmental progression in the structure of the instrument, during the formative research we did not group or order the items according to the developmental progression, instead placing them in random order. We further planned to administer the instrument to children individually, following a scripted protocol, but allowing for clinical interviewing techniques as deemed important.

Using these methods, we conducted formative research in several settings. First, we administered it ourselves to several young children, in conjunction with our development of the items. Second, we involved teachers, because, to be useful in the manner we planned for the Building Blocks project, any hypothetical learning trajectory should "speak" to practitioners as well as researchers. Therefore, we engaged a group of eight volunteers (from all the local teachers involved in the Building Blocks project) in a 6-month, iterative pilot test of the instrument. The goals included evaluating the appropriateness for children of different ages (e.g., does the wording make sense to your children) and for the teachers themselves (e.g., does the developmental progression make sense, do the items assess their thinking, does the interview help you learn about your children?), as well as informing us all on the nature of children's thinking, including assessing the validity of the developmental progression in the light of our observations of that thinking.

On the basis of the first pilot tests, we revised several items. For example, one teacher wrote, "Some sort of demo is needed to explain 'make different hexagons.'" Others agreed, and a demonstration was added. Similarly, another item that said, "Use four triangles such as these to make as many different shapes as you can," provided four triangle frames. Several participants reported that children, especially the youngest, just moved the manipulative shapes provided onto those pictured frames, interfering with their engagement and performance of the task (e.g., "At first L. picked up the triangles and filled the outlines at top. I pushed them into the open section and repeated the task"). Thus, we eliminated those frames in the next version of this and several similar items.

Once the items stabilized, both researchers and teachers interviewed 5 to 10 children each and wrote case studies of each (the reports, including a transcript, with analyses inserted, and a summary, were usually five to six single-spaced pages for each of 56 children, ages four to eight, $M = 6.6$ years). Each participant also completed a scoring sheet to have a consistent way to present the data to the group. To create the sheet, we reordered the items according to the developmental progression in the following way. First, we calculated an average score for each item displaying the order from most successful to least successful. Second, we analyzed each item in terms of the cognitive operations children were employing to solve them. On these bases, items were assigned to one or more levels of the developmental progression and ordered within these levels. We then created a color-coding scheme to differentiate the level of achievement within each level on each item. Each participant then reflected on her children's performance based on their observation notes and the corresponding scoring sheet. For example, visual inspection of a sheet provided evidence regarding the validity of the developmental progression in that items lower in the progression should, of course, be achieved before items later in the progression, and there should be few gaps or inconsistencies (e.g., if a child evinced success on all items from level n, followed by, say, 40% success on level $n + 1$, there should not then be substantial success on items at level $n + 2$).

Based on these analyses and on discussions of children's responses on individual items, we deleted or altered all items that were problematic. For example, we removed two items that were not appropriate to the age group (e.g., items in which knowledge of shape names such as pentagon or hierarchical inclusion of square and rectangle affected some children's responses). Other items were edited for ease of use or to maximize their potential for accurately assessing children's level of thinking. In addition to the modification of items, it was also decided that the instrument did not have sufficient items to evince behaviors characteristic of the higher levels of the developmental progression. It was also noted that few instances of those levels had been observed on the select items that offered the opportunity for such thinking. This was not surprising given the difference between ages of the children involved in the present study and the ages of the children who had responded at those levels in previous research. Thus, the decision was made to focus on the first four levels of the developmental progression in the summative research, but to note any behaviors related to the higher levels.

Further, these analyses indicated that the developmental progression appeared to be valid. All participants believed that the items that were retained constituted valid assessment of the kind of thinking described by each level of the developmental progression. In addition, all participants indicated that they believed that they could reliably classify children as exhibiting thinking on the developmental progression; however, the majority believed that one or more of the children they interviewed showed signs of thinking on each of two consecutive levels. Finally, the scoring sheets showed few gaps or inconsistencies. A numerical summary was calculated as follows: Each child whose average score for each level n was greater than or equal to the average score for level $n + 1$ was considered consistent with the developmental progression. With all items considered, 78% of the children were consistent. With the rejected items omitted, 84% were consistent.

Thus, at this point in our formative research and development, we had produced what we had reason to believe was a valid developmental progression and a valid instrument to assess children's thinking in terms of this developmental progression. Next, we conducted a large-scale, hypo-deductive study to evaluate the developmental progression.

SUMMATIVE RESEARCH

The formative research left us with the need to evaluate the validity of the developmental progression of shape composition and decomposition competencies with a large number of children across multiple grade levels and determine whether these levels do indeed form a disjunctive scale, that is, the behaviors characteristic of the higher levels successively replace those behaviors typical of the lower levels. To achieve this goal, we conducted a hypo-deductive, summative study administering the revised instrument to a large number of children.

Participants

This study included 72 children selected at random from all of the children who completed a participants permission letter. Eighteen children were selected from each grade level of prekindergarten through second grade (mean ages 4.1, 5.4, 6.4, and 7.4, respectively), drawn from six classrooms. The classrooms were selected from a group of teachers who expressed interest in participating in the research and who agreed not to teach any lessons related to shape composition and decomposition prior to completion of the research. Furthermore, to ensure children came from diverse settings and varied in socioeconomic status, classrooms were selected that were located in urban and suburban school districts.

Procedure

All children were interviewed individually by one of two graduate research assistants following the instrument protocol. The researchers also asked questions of the children, as in a clinical interview, whenever they believed that such questions would clarify the nature of children's thinking. Each session was videotaped. These tapes were partially transcribed, coded, and analyzed, both to complete the scoring of the instrument for each child and to identify additional data that could be used to refine the definitions of the levels. The two graduate assistants initially administered the protocol jointly to develop consistency. Additionally, a subset of the children were randomly selected and independently coded; the resulting inter-rater reliability was .88.

Instrument and Scoring

The instrument contained 17 items that, based on the pilot work, would best elicit responses reflective of the children's cognitive abilities in composition and, thus, the behaviors desired by the researchers to identify the level of the developmental progression at which the child is working. We modified the rubrics that were previously developed for each item to specifically indicate those behaviors indicative of each of the levels. This alternative analysis required a fundamental change from the way the items initially were scored in the pilot and provided a way for the items to be scored across the levels, with the dominant level and less dominant levels indicated for each item. For example, behaviors such as the inability to concatenate pieces to fill complex frames, concatenating pieces though the pieces might not share attributes of the frame, acceptance or rejection of shapes through placement of a shape on the frame, or the ability to use a mental image to select shapes were some of the behaviors monitored to score the four levels of the developmental progression on Item 1 (the "puzzle man"). Based on the frequency of occurrence of behaviors reflecting each level of the developmental progression (Precomposer to Shape Composer), a frequency rating of 0, 1, 2, or 3 was assigned. A 3 rating indicated that the behaviors exhibited on the item were dominated by that level, a 2 rating indicated several instances of behaviors reflective of that level were observed, a 1 rating indicated a single instance or minimal for that level, and a 0 rating indicated no behaviors reflective of that level were observed on that item. A score for each item was computed by summing the product of the frequency rating and level number. For example, say a child's score for Precomposer (level number 0), Piece Assembler, Picture Maker, and Shape Composer (level number 3), respectively, were for Item 1: 0, 2, 1, and 0, and for Item 2: 0, 3, 0, 0. The child's score for Item 1 would be $(0 \times 0) + (2 \times 1) + (1 \times 2) + (0 \times 3) = 4$ and for Item 2, $(3 \times 1) = 3$. The total score for the child was the sum of these item scores. In addition, to assign a dominant level in which each child was working, we summed the scores for each composition level for each child. Two researchers considered these scores for each child and independently assigned the child to operating at the dominant level or "in transition" between two contiguous levels. There were only two disagreements, which were resolved after noting that an algorithm could be applied which unambiguously classified children: If the greatest level score is at least twice the score of any contiguous level (and successively), then the child is assigned to that level. If it is twice the score of one contiguous level (e.g., $n - 1$) but not the other (e.g., $n + 1$) then the child is in transition (e.g., from level n to level $n + 1$). Any other case (e.g., three contiguous approximately equal level scores) would be considered unclassifiable and constituting disconfirming data. No such cases were encountered.

RESULTS

Findings generally supported the hypothesis that children demonstrate the various levels of thinking when given tasks involving the composition and decomposition of two-dimensional figures, and that older children and those with previous experience in geometry tend to evince higher levels of thinking. We briefly summarize the qualitative results (for a full report, see Wilson, 2002, as well as a work in progress on those data), and then describe the quantitative results.

Qualitative Results

Levels of thinking could be reliably differentiated, and children could be reliably assigned to a level of development (including those in transition, or in the process of developing the next level). For example, Mary, a preschool (4.3 years) child, exhibited behaviors that typified the behaviors of our early level of composition: the Piece Assembler. When Mary began working on the puzzle man item, she tried (correctly) to place a trapezoid in the foot as shown in Fig. 2a. The trapezoid was 180 degrees opposite of the orientation needed to fill the frame and Mary, through her minor rotations in each direction, was unable to arrive at the requisite orientation and thus rejected the piece. Moments later, she used the trapezoid to fill the arm, this time successfully rotating the shape to match the frame. After similarly filling the other arm, Mary returned to the legs and concatenated four squares to (incorrectly) cover one leg and two to cover the other before deciding to move on to another item (Fig. 2b). Mary's placement of the squares reflects her inability to attend to angle. It also reflects her transition from Precomposer as she is just recognizing that shapes can be concatenated to cover a region.

Kevin, a first-grade (6.0 years) child, demonstrated the behaviors used frequently by children at the Picture Maker level. The "picking and discarding" strategy that typifies this level was observed as Kevin repeatedly attempted to fill the dog puzzle item. As Kevin tried to fill the puzzle, shapes were selected and "tried out" for a fit through placement and manipulation of the shape directly on the puzzle; there was a notable lack of the construction of a mental image of the shape and its relation to the puzzle frame. Figure 2c displays the puzzle nearly completed.

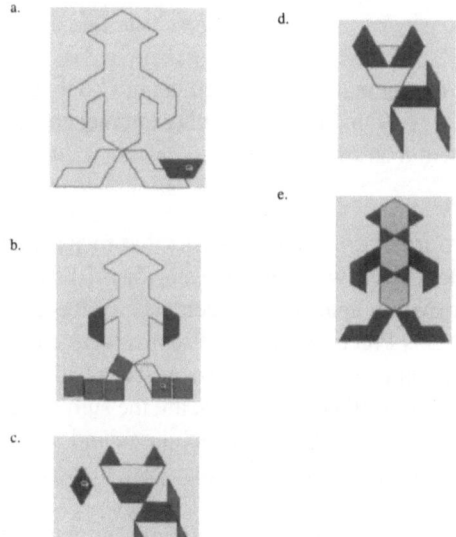

FIGURE 2 Examples of children's work on shape puzzles.

Kevin attempted to fit a rhombus into the open space, then a square, and eventually, unable to fill the open complex frame, he rejected the arrangement and cleared away the shapes in the head. He then placed a trapezoid along the left side of the head and similarly on the right, thereby creating two simple frames that did allow him to complete the puzzle (Fig. 2d).

The Picture Maker level of thinking evinced by Kevin precedes the Shape Composer level that is exemplified in the work of Alice, a second-grade (7.4 years) child. In this level we no longer observe the random selection of pieces, but rather deliberate selections are made as the child creates a mental image of how the shape can fill the frame. In addition, the process of completing a puzzle often becomes systematized. As Alice worked on the puzzle man she first placed a trapezoid on one arm, then the other, followed by a rhombus on each arm. Similarly, Alice carefully considered how to fill the leg as she looked back and forth from the shapes to the puzzle prior to making her selection. Once she filled one leg with two trapezoids, she was able to think of the concatenated pieces as a whole and simply duplicated the process on the other leg. She filled the body of the puzzle man in a similarly systematic way with the finished puzzle reflecting this (Fig. 2e).

Across all 72 children, examination of the items indicating attainment of each level similarly confirmed the hypotheses. If a child evinced a level of thinking on one item, they were more likely than not to attain it on the other items measuring that level. Most exceptions involved the highest level the child had attained; many children were in the process of developing that level of thinking. Quantitative summaries provided additional support for these conclusions.

Quantitative Results

Reliability. The reliability of the instrument was assessed using Cronbach's alpha. The alpha score was calculated to be .945, indicating a high internal consistency among the items on the assessment instrument.

Validity of the developmental progression. The analyses that focused on the developmental progression, and establishing whether the levels form a disjunctive scale, began by completing a correlation analysis between each of the children's level scores. Thus, all 72 children's Precomposer level scores were correlated with all 72 children's Piece Assembler level scores, and so on. The results are presented in Table 2. The level scores, as described previously, are the sum of occurrences of the behavior dominance of each level, within each item, and multiplied by a factor of 0, 1, 2, or 3 reflecting the dominance of that level in the children's thinking.

It was expected that the correlation matrix would display a simplex structure if indeed the levels form a disjunctive scale. That is, as one reads the table, beginning on the main diagonal of any row, the correlations should steadily decrease as one moves away from the diagonal in either direction (Davison, Robbins, & Swanson, 1978). We would expect a child who scores high in a given level would be likely to score higher on the levels immediately adjacent to the given level, than on the levels nonadjacent to the given level, thus resulting in higher correlations between adjacent levels than between nonadjacent levels.

The simplex structure is evident in Table 2, with one exception. The pattern is not consistent with the correlation between the Shape Composer level and the Precomposer level. It is interesting that this correlation is less negative than the previous correlation between the Picture Maker level and the Precomposer level, because one would expect the lowest level and the highest level of the developmental progression to be nearly exclusive of one another (i.e., highly negatively correlated). To understand why the correlation is not what was expected, consider the plots of the individual level scores. Figure 3 displays the Picture Maker level versus Precomposer level scores and depicts, as expected, a pattern that reflects the expectation that children who score higher on the Picture Maker level tend to score

TABLE 2
Level Score Correlations

	Precomposer	Piece Assembler	Picture Maker	Shape Composer
Precomposer	1	.198	−.659	−.405
Piece Assembler	.198	1	−.551	−.699
Picture Maker	−.659	−.551	1	.348
Shape Composer	−.405	−.699	.348	1

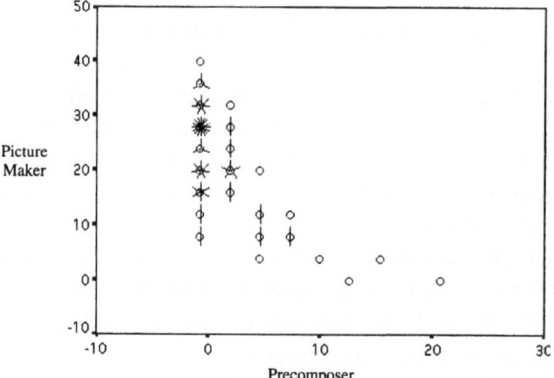

FIGURE 3 Scatterplot of Precomposer and Picture Maker level scores. Multiple identical data points are represented by radial line segments emanating from the center points.

very low or zero on the Precomposer level. The dominant pattern observed in the plot is the large number of points that lie on a vertical line associated with a score of zero on the Precomposer level, and thus, most children who scored high on the Picture Maker level, appropriately scored a zero on the Precomposer level. Similar results are shown in Figure 4 for the Shape Composer level versus the Precomposer level scores.

The pattern observed in Figure 3 of points predominantly falling on lines associated with scores of zero on one of the levels is more prevalent in this plot. This again was expected. That is, children who exhibit behaviors typical of Shape Composers do not exhibit behaviors associated with Precomposers, and vice versa. This relation verbally describes a highly negative correlation; however, the plot displays

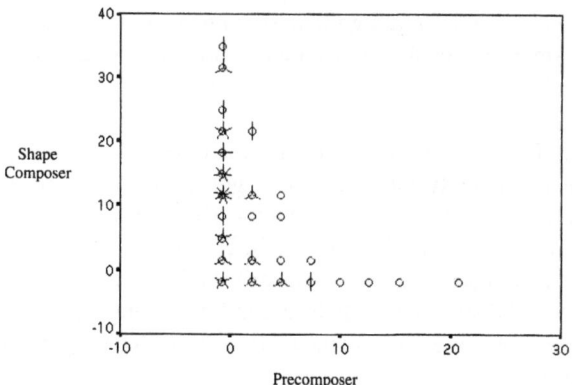

FIGURE 4 Scatterplot of Precomposer and Shape Composer level scores.

an *L* shape, rather than the highly negative pattern of points clustered about a line with a slope of negative one. This *L* form is the result of the scoring method employed in the analysis. Children who score higher on the Shape Composer level would be expected to score lower on the Precomposer level, but actually can only score as low as a zero. This causes a floor effect and the resulting *L* plot. In fact, if children's scores on the two levels on the ends of the developmental scale were to follow the developmental progression precisely, we would have points only on the *L*, indicating the exclusive nature of these two levels. Thus, the correlation between these two levels that we would hope to see in the matrix, when calculated under this scoring scheme, would be the correlation of points forming an *L*, which is negative .5. With this in mind, the correlation value of –.405 between the Precomposer level and the Shape Composer level, although breaking the simplex pattern, actually is consistent with the predictions of the hypothesized developmental progression. Other levels that are closer on the developmental progression would exhibit a larger negative correlation under this scoring scheme.

Thus, the correlation matrix, although not displaying a perfect simplex structure, does reflect precisely what this scoring scheme permits: a quasi-simplex structure with the exceptions actually reflecting the structure of the scoring scheme employed in the analysis and thus consistent with the hypothesis.

Levels of thinking: Development from prekindergarten to second grade. Another test is whether the progression shows development across ages. Means (and standard deviations) for children's total scores were 4.07 (.40), 5.39 (.28), 6.43 (.25), 7.41 (.28) for grades prekindergarten, kindergarten, first, and second, respectively. A one-way analysis of variance indicated that they differed significantly, $F (3, 68) = 394.87$, $p < .001$. Post hoc, pairwise comparisons between adjacent grades were made using an independent samples *t*-test and the Bonferroni correction procedure. The results were significant ($p < .001$) for all comparisons. That is, kindergarteners scored significantly higher than prekindergartners, first graders scored significantly higher than kindergartners, and second graders scored significantly higher than first graders. Thus, findings provide support for the developmental nature of our progression.

Given the multiple forms for support for the developmental progression, it is valid to ask the theoretically and practically significant question regarding the distribution of levels for children of different ages. The level scores, as previously described, allowed the children's behaviors to be categorized as indicative of a single level or as in transition between levels. The distribution of levels is presented in Table 3. The pattern of decreasing early levels and increasing higher levels of development as age increases is evident in the table along with the significant number of children in transition.

Although these findings can be interpreted as indicating that children simply score higher as they mature, we do know that these children's teachers were pro-

TABLE 3
Distribution of Primary Level Scores and Transitional Level Scores
Across Grades

Level	Grade				
	Pre-K	K	1st	2nd	Total
Precomposer	1				1
Precomposer/Piece Assembler	3				3
Piece Assembler	5	2		1	8
Piece Assembler/Picture Maker	9	3	1	1	14
Picture Maker		9	10	2	21
Picture Maker/ Shape Composer		4	6	13	23
Shape Composer			1	1	2
Total	18	18	18	18	72

viding experiences with geometry. The role of these experiences needs to be studied, especially in promoting growth to the shape composer and higher levels (van
Hiele, 1986).

DISCUSSION AND IMPLICATIONS

The ability to describe, use, and visualize the effects of composing and decomposing geometric shapes is a significant aspect of geometric thinking. To better understand the development of this ability, we designed a hypothetical learning trajectory, including a developmental progression based on previous research, as well as
an instrument to assess levels of this progression. We tested both the progression
and the instrument through a series of investigations.

Findings provide strong support for the first four levels of the developmental
progression (evidence supported all, or at least did not weigh against, any levels,
but data are insufficient to make firm claims regarding the top three levels). First,
the original hypothetical learning trajectory and the developmental progression
underlying it emerged from naturalistic observations of young children composing
shapes, both our own (Clements et al., 1997; Sarama et al., 1996, Sarama &
Clements, in preparation) and those of other researchers (Mansfield & Scott, 1990;
Sales, 1994). Second, the levels of the developmental progression were tested
iteratively in formative research that involved researchers and teachers. Their case
studies indicated that about four-fifths of the children studied evinced behaviors
consistent with the developmental progression (using an early version of the instrument). By the end of this phase, all participants believed the developmental
progression and the items retained to measure levels in the progression to be valid
and that they could reliably classify children as exhibiting thinking on the progres-

sion. Third, a summative study employed the final instrument with 72 randomly selected children from prekindergarten to second grade. Analyses revealed that the level scores fit the hypothesized structure in which scores from one level would be more highly correlated with scores immediately adjacent to that level than to scores on levels nonadjacent to the given level. Further, the developmental progression showed development across ages, with children at each grade scoring significantly higher than those at the preceding grade. Based on these results, we conclude that we have produced a valid developmental progression.

These findings make three contributions to the research literature. First, they extend theories and research in children's learning of geometry (Clements, 2004) to include the development of shape composition, providing a more detailed description of this development than any of which we are aware. In the developmental progression underlying our hypothetical learning trajectory, children move through several distinct levels of thinking and competence in the domain of composition and decomposition of geometric figures (as described in Table 1). These competencies include the child's ability to make pictures or designs by combining shapes (initially by trial and error, then by considering attributes, and also progressing to leaving fewer gaps to full tiling at the upper levels); create, maintain, and operate on a shape as a mathematical unit with measurable attributes; and compose two or more other shapes, eventually applying a uniting operation to reconceptualize the composite shape as a new unit that is conceptualized as independent entity. Consistent with the research of Piaget and Inhelder (1967), these processes operate first on physical shapes (figurative knowledge) and later on mental constructs (i.e., in mental imagery and eventually as explicit mathematical entities, or operative knowledge).

Precomposers have almost none of these competencies, usually resisting shape composition. Piece Assemblers concatenate shapes to form pictures, but maintain semantic separation of the shapes and have limited ability to use turns or flips. Picture Makers understand and use turns and occasionally flips to concatenate shapes to form pictures in which multiple shapes play a single role, often using simple geometric attributes (e.g., side length), but they use trial and error in placing shapes contiguously and do not conceptualize the resulting shape as a new unit. Shape Composers can operate mentally; this accounts for their choice of shapes based on properties, use of geometric motions with intentionality and anticipation, and ability to create a new geometric unit—the composite shape (cf. Vurpillot's, 1976, "secondary level"). For the children in our summative study, preschoolers tended to be Piece Assemblers or in transition between the Piece Assembler and Picture Maker levels, kindergartners were mostly Picture Makers, first graders were Picture Makers or in transition between the Picture Maker and Shape Composer levels, and second graders were in transition between the Picture Maker and Shape Composer levels. Recall that this is without any specially designed curriculum intervention.

Second, the instrument is a contribution to the research literature. Results indicate that the instrument is a reliable and valid assessment of shape composition competencies, which could be useful to others conducting research in the domain.

This work's third and final contribution to the literature lies in its support for the use of developmental progressions as a foundation for hypothesized learning trajectories, adding to the nascent literature on this idea. Although we reported on several studies, space does not permit us to describe the coincident development and revision of the entire hypothetical learning trajectory. In brief, throughout the development of activities for the Building Blocks project, we continued formative research that included shape composition, conducting small group and whole class teaching experiments that provided qualitative data on the hypothetical learning trajectory (see Sarama & Clements, 2002, and the examples of activities in Figs. 5 and 6). These data influenced our thinking throughout the studies discussed in this report. In addition, we conducted summary research (Clements & Sarama, 2003c) that indicated gains of more than one standard deviation in geometry; the highest gains were in the area of shape composition. In combination, these findings indicate that such developmental progressions and learning trajectories allow the encapsulation of significant cognitive constructions within a sequence that is relevant to curriculum development and teaching (Clements, 2002; Cobb & McClain, 2002; Gravemeijer, 1999; Sarama & Clements, in press; Simon, 1995). The results lend credence to our argument (Clements, 2002; Sarama & Clements, in press) that existing research should be a primary means of constructing the first draft of these learning trajectories. We believe that more widespread adoption of this approach could help ameliorate the difficulty many development teams appear to have incorporating the research of others.

The findings also make three contributions to educational practice. First, the instrument could be adapted by teachers to assess children's entering level of thinking in shape composition, as well as their development of these levels throughout the school year.

FIGURE 5 Child working with physical pattern blocks, solving puzzles at their level. The *DLM Early Childhood Express Math Resource Guide* (Clements & Sarama, 2003a) includes numerous puzzles for each level in the learning trajectory.

a.

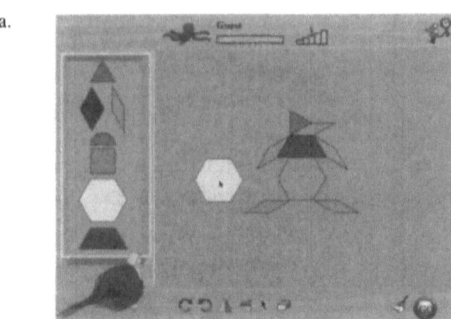

b.

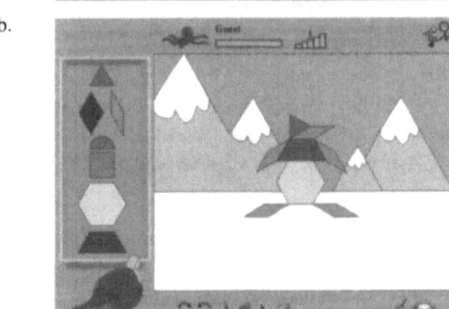

FIGURE 6 The *DLM Early Childhood Math Software*, which accompanied the *Resource Guide*, automatically provides puzzles at the children's current level of thinking. (a) A puzzle in progress. (b) Feedback after a successful solution.

Second, the developmental progression (as well as the other components of the hypothetical learning trajectory) was created in collaboration with teachers, and the description, assessment, and even labeling of the levels was designed to maintain its comprehensibility. This is significant because there is evidence that successful teachers strive to help children move not through a curriculum, but through learning trajectories (e.g., Fuson, Carroll, & Drueck, 2000). Putting learning trajectories at the center helps all parties connect goals, curriculum components, assessment, and teaching strategies. It facilitates teachers' learning about math, how children think about and learn this math, and how such learning is supported by the curriculum and its teaching strategies by illuminating potential developmental paths (Ball & Cohen, 1999) and thus brings coherence and consistency to math goals and curricula, and assessments.

Third, the developmental progression should, of course (given its initial purpose), be useful for curriculum and software design. As mentioned, we have built our shape composition trajectory into the Building Blocks software (e.g., Clements & Sarama, 2003a). In activities such as "Shape Puzzles," children use mathematical actions-on-objects to compose geometric shapes. The shapes are available in one of several different sets (e.g., pattern blocks, tangrams); the actions include duplication, translation, rotation, and reflection. The software assesses children's performance completing puzzles (not unlike those in the figures for this article) and moves them up (or down) through the trajectory by providing puzzles appropriate to each level of thinking. For example, the precomposing child is provided puzzles in which a frame suggests the placement of each shape; further, each shape plays a separate role in the picture. For the child who is at the Piece Assembler level, the software encourages development of thinking at the Picture Maker level by providing puzzles that suggest the placement of shapes, but in which several shapes together can play a single semantic role in the picture. To encourage Shape Composer thinking, the software provides puzzles in which the child must completely fill a region that consists of multiple corners, requiring selecting and placing shapes to match angles. Teachers also encourage free-form composition of shapes and encourage children to discuss their work on all tasks. The aforementioned summary research (Clements & Sarama, 2003c) provides evidence of this curriculum's effectiveness.

Although this research has made several contributions, there are several lines of inquiries that should be pursued before the theory can be considered fully supported. First, research should address the highest three levels of the developmental progression, which were inadequately assessed by the present data. Second, we have not investigated whether children evince the same levels of thinking working with puzzles versus free-form shape composition tasks. Most of the present research used the former format, so results and conclusions may be more reliable for puzzle tasks. Third, full evaluation of any learning trajectory should include tracing development longitudinally. We are presently pursuing the two latter lines of research. For example, we are assessing children longitudinally using teaching experiments to ascertain if children do indeed develop through each level (e.g., with efficacious instruction, will some children skip one or more levels). This also will help determine whether the levels are linked primarily to maturity, or whether a sequence of activities aligned with the learning trajectory would promote the children's development within the domain of composition and decomposition of geometric figures. As previously noted, this domain corresponds with, and supports, children's ability to compose and decompose numbers (Clements et al., 1997). This suggests another objective the longitudinal study should consider. If indeed specific instructional activities do promote the children's progression in shape composition and decomposition, then a corresponding development within the domain of number composition and decomposition might be measured as well.

ACKNOWLEDGMENTS

This article was supported in part by the National Science Foundation under Grants ESI–9730804, "Building Blocks—Foundations for Mathematical Thinking, Pre-Kindergarten to Grade 2: Research-based Materials Development" (Douglas H. Clements and Julie Sarama, Co-PIs) and REC–9903409, "Technology-Enhanced Learning of Geometry in Elementary Schools" (Daniel Watt, Douglas H. Clements, and Richard Lehrer, Co-PIs), and by the Interagency Educational Research Initiative (NSF, DOE, and NICHHD) Grant REC–0228440, "Scaling Up the Implementation of a Pre-Kindergarten Mathematics Curricula: Teaching for Understanding with Trajectories and Technologies"(D. H. Clements, J. Sarama, A. Klein, & P. Starkey, co-PIs). Any opinions, findings, and conclusions or recommendations expressed in this material are those of the author and do not necessarily reflect the views of the National Science Foundation.

We thank Daniel Watt, Richard Lehrer, and Dick Lesh for their advice during the conduct of this research.

REFERENCES

Ball, D. L., & Cohen, D. K. (1999). *Instruction, capacity, and improvement.* (CPRE Research Report No. RR–043). Philadelphia: University of Pennsylvania, Consortium for Policy Research in Education.

Clements, D. H. (2002). Linking research and curriculum development. In L. D. English (Ed.), *Handbook of international research in mathematics education* (pp. 599–630). Mahwah, NJ: Lawrence Erlbaum Associates, Inc.

Clements, D. H. (2004). Geometric and spatial thinking in early childhood education. In D. H. Clements, J. Sarama, & A.-M. DiBiase (Eds.), *Engaging young children in mathematics: Standards for early childhood mathematics education* (pp. 267–297). Mahwah, NJ: Lawrence Erlbaum Associates, Inc.

Clements, D. H., & Battista, M. T. (2000). Designing effective software. In A. E. Kelly & R. A. Lesh (Eds.), *Handbook of research design in mathematics and science education* (pp. 761–776). Mahwah, NJ: Lawrence Erlbaum Associates, Inc.

Clements, D. H., Battista, M. T., & Sarama, J. (2001). Logo and geometry. *Journal for Research in Mathematics Education Monograph Series, 10.* Reston, VA: National Council of Teachers of Mathematics.

Clements, D. H., Battista, M. T., Sarama, J., & Swaminathan, S. (1997). Development of students' spatial thinking in a unit on geometric motions and area. *The Elementary School Journal, 98,* 171–186.

Clements, D. H., & Sarama, J. (1998). Shapes—Making Shapes [Computer software]. Palo Alto, CA: Dale Seymour.

Clements, D. H., & Sarama, J. (2003a). *DLM Early Childhood Express Math Resource Guide.* Columbus, OH: SRA/McGraw-Hill.

Clements, D. H., & Sarama, J. (2003b). DLM Math Software [Computer software]. Columbus, OH: SRA/McGraw-Hill.

Clements, D. H., & Sarama, J. (2003c). *Effects of a preschool mathematics curriculum: Summary research on the Building Blocks project.* Manuscript submitted for publication.

Clements, D. H., Sarama, J., Battista, M. T., & Swaminathan, S. (1996). Development of students' spatial thinking in a curriculum unit on geometric motions and area. In E. Jakubowski, D. Watkins, & H. Biske (Eds.), *Proceedings of the eighteenth annual meeting of the North America Chapter of the International Group for the Psychology of Mathematics Education* (Vol. 1, pp. 217–222). Columbus, OH: ERIC Clearinghouse for Science, Mathematics, and Environmental Education.

Cobb, P., & McClain, K. (2002). Supporting students' learning of significant mathematical ideas. In G. Wells & G. Claxton (Eds.), *Learning for life in the 21st Century: Sociocultural perspectives on the future of education.* New York: Cambridge University Press.

Davison, M. L., Robbins, S., & Swanson, D. B. (1978). Stage structure in objective moral judgments. *Developmental Psychology, 14,* 137–146.

Fuson, K. C., Carroll, W. M., & Drueck, J. V. (2000). Achievement results for second and third graders using the *Standards*-based curriculum *Everyday Mathematics. Journal for Research in Mathematics Education, 31,* 277–295.

Gravemeijer, K. P. E. (1994). Educational development and developmental research in mathematics education. *Journal for Research in Mathematics Education, 25,* 443–471.

Gravemeijer, K. P. E. (1999). How emergent models may foster the constitution of formal mathematics. *Mathematical Thinking and Learning, 1,* 155–177.

James, W. (1958). *Talks to teachers on psychology: And to students on some of life's ideas.* New York: Norton.

Lane, S., & Rosser, R. A. (1987, April). *Validation of young children's geometric skill sequences.* Paper presented at the meeting of the American Educational Research Association, Washington, DC.

Mansfield, H. M., & Scott, J. (1990). Young children solving spatial problems. In G. Booker, P. Cobb, & T. N. deMendicuti (Eds.), *Proceedings of the 14th Annual Conference of the Internation Group for the Psychology of Mathematics Education* (Vol. 2, pp. 275–282). Oaxlepec, Mexico: International Group for the Psychology of Mathematics Education.

Piaget, J., & Inhelder, B. (1967). *The child's conception of space* (F. J. Langdon & J. L. Lunzer, Trans.). New York: W. W. Norton.

Reynolds, A., & Wheatley, G. (1996). Elementary students' construction and coordination of units in an area setting. *Journal for Research in Mathematics Education, 27,* 564–581.

Sales, C. (1994). *A constructivist instructional project on developing geometric problem solving abilities using pattern blocks and tangrams with young children.* Unpublished master's thesis, University of Northern Iowa, Cedar Falls.

Sarama, J. (2004). Technology in early childhood mathematics: *Building Blocks™* as an innovative technology-based curriculum. In D. H. Clements, J. Sarama, & A.-M. DiBiase (Eds.), *Engaging young children in mathematics: Standards for early childhood mathematics education* (pp. 361–375). Mahwah, NJ: Lawrence Erlbaum Associates, Inc.

Sarama, J., & Clements, D. H. (2002). *Building Blocks* for young children's mathematical development. *Journal of Educational Computing Research, 27*(1&2), 93–110.

Sarama, J., & Clements, D. H. (in press). Linking research and software development. In K. Heid & G. Blume (Eds.), *Technology in the learning and teaching of mathematics: Syntheses and perspectives.* New York: Information Age Publishing, Inc.

Sarama, J., Clements, D. H., & Vukelic, E. B. (1996). The role of a computer manipulative in fostering specific psychological/mathematical processes. In E. Jakubowski, D. Watkins, & H. Biske (Eds.), *Proceedings of the Eighteenth Annual Meeting of the North America Chapter of the International Group for the Psychology of Mathematics Education* (Vol. 2, pp. 567–572). Columbus, OH: ERIC Clearinghouse for Science, Mathematics, and Environmental Education.

Simon, M. A. (1995). Reconstructing mathematics pedagogy from a constructivist perspective. *Journal for Research in Mathematics Education, 26,* 114–145.

Steffe, L. P., & Cobb, P. (1988). *Construction of arithmetical meanings and strategies.* New York: Springer-Verlag.

van Hiele, P. M. (1986). *Structure and insight: A theory of mathematics education.* Orlando, FL: Academic.

Vurpillot, E. (1976). *The visual world of the child.* New York: International Universities Press.

Wilson, D. C. (2002). *Young children's composition of geometric figures: A learning trajectory.* Unpublished doctoral dissertation, University of Buffalo, State University of New York.

MATHEMATICAL THINKING AND LEARNING, 6(2), 185–204

Applying Cognition-Based Assessment to Elementary School Students' Development of Understanding of Area and Volume Measurement

Michael T. Battista

Department of Teacher Education
Michigan State University

As part of a discussion of cognition-based assessment (CBA) for elementary school mathematics, I describe assessment tasks for area and volume measurement and a research-based conceptual framework for interpreting students' reasoning on these tasks. At the core of this conceptual framework is the notion of levels of sophistication. I provide details on an integrated set of levels for area and volume measurement that (a) starts with the informal, preinstructional reasoning typically possessed by students, (b) ends with the formal mathematical concepts targeted by instruction, and (c) indicates cognitive plateaus reached by students in moving from (a) to (b).

Because cognition is the core substance of understanding and sense making, cognition-based assessment (CBA) is critical to understanding, monitoring, and guiding students' development of powerful mathematical thinking. The CBA project is applying the results, theories, and methods of modern research in mathematics education to create an assessment system that can be used to assess in detail the cognitive underpinnings of the progress students make in developing understanding and mastery of elementary school mathematics (Battista, 2001b). This article first outlines the theoretical rationale for CBA and then describes an example that illustrates the nature of such assessment.

Requests for reprints should be sent to Michael T. Battista, Michigan State University, East Lansing, MI 48824. E-mail: mbattist@msu.edu

THEORETICAL RATIONALE

Learning and Instruction

According to the psychological constructivist view of how students learn mathematics with understanding, the way students construct, interpret, think about, and make sense of mathematical ideas is determined by the elements and organization of the relevant mental structures that the students are currently using to process their mathematical worlds (e.g., Battista & Larson, 1994; Bransford, Brown, & Cocking, 1999; De Corte, Greer, & Verschaffel, 1996; Goldin, 1992; Greeno, Collins, & Resnick, 1996; Hiebert & Carpenter, 1992; Lester, 1994; National Research Council, 1989; Romberg, 1992; Schoenfeld, 1994; Steffe & Kieren,1994). To construct new knowledge and make sense of novel situations, students build on and revise their current mental structures through the processes of action, reflection, and abstraction.

Action is conscious, physical or imagined, purposeful activity. Examples include moving—or visualizing the movement of—objects, manipulating symbols, creating or manipulating (say, electronically) drawings, and imagining the construction or transformation of figures. *Abstraction* is the process by which the mind selects, coordinates, unifies, and registers in memory a collection of mental items or acts that appear in the attentional field (more will be said about this process later). *Reflection* is the conscious process of re-presenting actions and experiences to consider their results or how they are composed (von Glasersfeld, 1995). *Reflective abstraction* takes mental operations performed on previously abstracted items as elements and coordinates them into new forms or structures, which, in turn, can become the content—what is acted on—in subsequent action, reflection, and abstraction.

A major component of the psychological constructivist view of mathematics learning and teaching is its attention to students' construction of meaning for specific mathematical topics (Cobb, Wood, & Yackel, 1990; Steffe & Kieren, 1994). For numerous mathematical topics, researchers have found that students' development of conceptualizations and reasoning can be characterized in terms of *levels of sophistication* (e.g. Battista & Clements, 1996; Battista, Clements, Arnoff, Battista, & Borrow, 1998; Cobb & Wheatley, 1988; Steffe, 1992; van Hiele, 1986). These levels lie at the heart of the CBA conceptual framework for understanding students' learning progress. A set of levels for a topic (a) starts with the informal, preinstructional reasoning typically possessed by students, (b) ends with the formal mathematical concepts targeted by instruction, and (c) indicates cognitive plateaus reached by students in moving from (a) to (b). These plateaus can be pictured as in Figure 1. The "cognitive terrain" that students must ascend to attain understanding of a particular major mathematical idea is multidimensional and includes several key processes and conceptualizations. However, even though these pro-

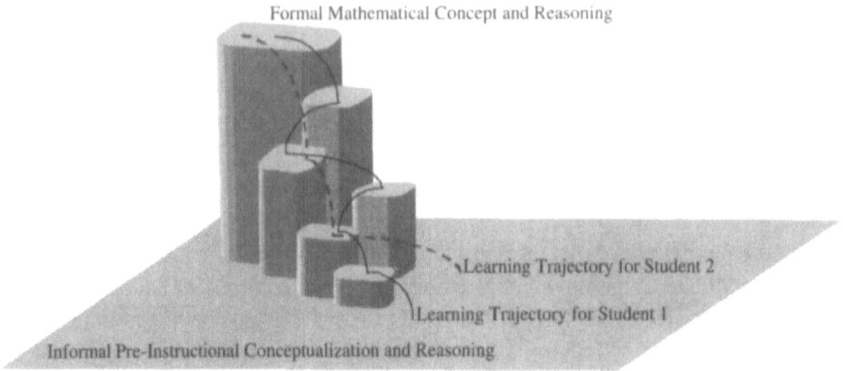

FIGURE 1   Levels of sophistication plateaus.

cesses can develop somewhat independently, there are identifiable levels of sophistication evidenced in students' development of understanding and competence with the idea.

A levels-model for a topic describes not only cognitive plateaus, but also what students can and cannot do, students' conceptualizations and reasoning, cognitive obstacles that obstruct learning progress, and mental processes needed both for functioning at a level and for progressing to higher levels. The levels are derived from analysis of both the mathematics to be learned and empirical research on students' learning of the topic.

The jumps in the ascending plateau structure of a levels-model represent cognitive restructurings evidenced by observable increases in sophistication in students' reasoning about a topic. Furthermore, an ideal levels-of-sophistication model for a topic provides indications of jumps in sophistication that are small enough to fall within students "zones of construction." That is, a student should be able to accomplish the jump from conceptualizing and reasoning at Level N to conceptualizing and reasoning at Level $N + 1$ by making a significant abstraction, in a particular context, while working to solve an appropriate problem or set of problems.[1]

However, because the levels are compilations of empirical observations of the thinking of many students and because students' learning backgrounds and mental processing differ, a particular student might not pass through every level for a topic; he or she might skip some levels or pass through them so quickly that the passage is difficult to detect. Even with this variability, however, the levels still describe the plateaus that students achieve in their development of reasoning about a

---

[1]The jump in reasoning may apply to restricted contexts, not to all contexts connected with the mathematical topic. That is, the jump can be tightly situated rather than global.

topic. They indicate major landmarks that research has shown students often pass through in "constructive itineraries" or learning trajectories for these topics. Thus, such levels provide an excellent conceptual framework for understanding the paths students travel to achieve meaningful learning of a topic.

The pedagogical importance of describing cognitive levels of sophistication for a mathematical topic resides in the fact that instruction that produces conceptual understanding and powerful reasoning for a mathematical topic must be firmly guided by detailed, research-based knowledge of the development of students' thinking about the topic (Carpenter & Fennema, 1991; Cobb et al., 1990; Steffe & D'Ambrosia, 1995). Selecting/creating instructional tasks, adapting instruction to students' needs, and assessing students' learning progress require detailed, cognition-based knowledge of how students construct meanings for the specific mathematical topics targeted by instruction.

Cognition-based Assessment

Assessment and instruction.   To implement mathematics instruction that genuinely and effectively supports students' construction of mathematical meaning and competence, teachers must not only understand cognition-based research on students' learning of particular topics, they must also be able to use that knowledge to determine and monitor the development of their own students' reasoning. Thus, teachers need (a) clear research-based descriptions of students' development of meaning for topics, and (b) sets of assessment tasks that enable them to determine how each of their own students' reasoning is progressing. CBA meets these needs when it includes the following three critical components.

1. *Descriptions of core mathematical ideas and reasoning processes.* Core ideas and reasoning processes enable powerful problem solving, novel application of concepts, explanation and justification of mathematical reasoning, and interconnection of ideas. They make it possible not only for students to do mathematics, but also to understand why what they do is valid. One example of a core mathematical idea is place value. Examples of core reasoning processes are spatial structuring and using composite units (see the example that follows).

2. *For each core idea, research-based descriptions of the cognitive constructions students must make in developing understanding of the idea.* These descriptions provide the conceptual frameworks necessary for understanding students' learning. They include levels of sophistication that students pass through in moving from their intuitive ideas and ways of reasoning to goal states of learning, cognitive obstacles that students face in learning, and fundamental mental processes that underlie concept development and reasoning for the idea.

3. *For each core idea, coherent sets of assessment items that enable educators to investigate students' cognitions and precisely locate students' positions in the*

constructive itineraries typically taken in acquiring competence with the idea. Some assessment items are appropriate for short one-on-one structured interviews, some for paper-and-pencil format. Some items assess what students can do, and some reveal students' reasoning and underlying mathematical cognitions.

The three components—identification of core ideas, conceptual frameworks for understanding students' reasoning about the ideas, and coherent sets of assessment tasks—are the critical components of an assessment system for understanding the development of students' mathematical reasoning.

Assessment and evaluation of instructional effectiveness.   There are two fundamentally different approaches to assessing student learning. Traditional assessment attempts to determine if students have acquired particular mathematical knowledge, skills, and types of reasoning. In contrast, CBA enables educators to investigate what students have learned. Using both the results and qualitative techniques of modern research on students' mathematics learning, CBA can carefully examine the exact nature of students' cognitions.

To guide teaching and curriculum development and to truly understand the effects of various types of instruction on students' mathematics learning, mathematics educators need both the *if* and *what* approaches to assessment. On one hand, we need to determine if students are acquiring specific knowledge and skills. On the other hand, we must determine the exact nature of students' ideas, reasoning, and cognitions as they attempt to learn particular mathematical topics.

Assessment and support for instructional improvement.   There are three ways in which CBA can serve as a powerful tool for encouraging and supporting teachers in their efforts to improve their mathematics teaching. First, to implement instruction of the kind envisioned in professional standards and consistent with modern research on mathematics learning, teachers must possess (a) a deep understanding of the mathematics to be taught, including the core ideas that guide powerful thinking relevant to that mathematics; (b) knowledge of how students at particular grade levels think about particular mathematics topics, including typical constructive itineraries taken by students on these topics; and (c) the ability to use this knowledge to analyze how the individual children in their classrooms are thinking about instructionally targeted mathematical topics (Bransford et al., 1999; Carpenter et al., 1999). CBA identifies and describes the core ideas, along with students' thinking about those ideas, and, through its assessment tasks, enables teachers to apply that knowledge to analyze their students' mathematical thinking. Research shows that such knowledge can improve students' learning. Indeed, as Fennema et al. (1996) stated, their research, "when viewed in conjunction with those of other studies, provide a convincing argument that one major way to improve mathematics instruction and learning is to help teachers understand the mathematical thought processes of their students" (p. 432).

Second, assessment that shows teachers that their students are not learning mathematics as deeply as teachers believe is a powerful motivator for reform. When teachers' use of CBA indicates to them that their students do not understand mathematics nearly as well as they assumed, a powerful impetus for instructional change occurs.

Third, and finally, because teachers are sent mixed signals by professional recommendations, students, parents, and high-stakes testing, they are often confused about goals, assessment, and even the very nature of the mathematics they should teach. They often choose instructional materials from various sources without a coherent plan. By helping teachers understand core ideas that students should learn, how these ideas develop, and how to assess them, CBA can provide clear instructional foci that teachers can use to improve their teaching. In fact, because one of the consequences of testing is to signal to students and teachers those aspects of learning that are valued, CBA can shift the instructional focus from student behavior to the cognition that underlies mathematical reasoning and performance.

AN EXAMPLE OF COGNITION BASED ASSESSMENT: UNDERSTANDING THE DEVELOPMENT OF STUDENTS' THINKING ABOUT MEASURING AREA AND VOLUME

To demonstrate the nature of CBA, I now discuss some assessment tasks for area and volume measurement and a conceptual framework for interpreting students' reasoning on these tasks. A central component of this conceptual framework is a research-based description of levels of sophistication in the development of students' reasoning about these topics.

Previous research studies have described separate levels of sophistication in students' reasoning about area and volume measurement (Battista, 1999; Battista et al., 1998; Battista & Clements, 1996). To engender a more parsimonious, coherent, and efficient analysis of students' thinking about area and volume, the CBA project has integrated the models for area and volume measurement into a more general, cognition-based conceptual framework.[2]

---

[2]The portion of this article dealing with two-dimensional and three-dimensional arrays is based in part on Battista (2003).

Underlying Mental Processes

A core idea in developing competence with measuring area and volume in standard measurement systems is understanding how meaningfully to enumerate arrays of squares and cubes such as those shown in Figures 2a and 2b.

As individuals purposely interact with the physical world and other people, their minds construct sets of cognitive structures that enable them to conceptualize, reason about, and manage those interactions (von Glasersfeld, 1995). Among the cognitive processes that enable such cognitive constructions, abstraction is critical (Steffe, 1988). At its most basic or *perceptual level*, abstraction isolates an item in the experiential flow and grasps it as a unit (von Glasersfeld, 1995). When material has been sufficiently abstracted so that it can be re-presented (re-created) in the absence of perceptual input, it has reached the *internalized level*. Material has reached the *interiorized level* when it has been disembedded from its original perceptual context and it can be freely operated on in imagination, including projecting it into other perceptual material and utilizing it in novel situations. Interiorization is "the most general form of abstraction; it leads to the isolation of structure (form), pattern (coordination), and operations (actions) from·experiential things and activities" (Steffe & Cobb, 1988, p. 337).

Four additional cognitive processes are essential for meaningful enumeration of arrays of squares and cubes and are used throughout this article to explain students' thinking: forming and using mental models, spatial structuring, units-locating, and organizing-by-composites. In the *forming and using mental models* process, individuals create and use imagistic or recall-of-experience-like mental representations that have structures isomorphic to the perceived structures of the situations they represent (Battista, 1999). Mental models consist of integrated sets of abstractions that are activated to visualize, comprehend, and reason about situations that one is dealing with in action or thought. For instance, to give someone travel directions without the use of a map, one reflects on a mental model of the locality to visualize and describe a route to follow. In the *spatial structuring* process, individuals abstract an object's composition and form by identifying, interrelating, and organizing its components. For instance, to spatially structure a rectangular array of squares, one might see it as a set of rows, one for each square in a column. Stu-

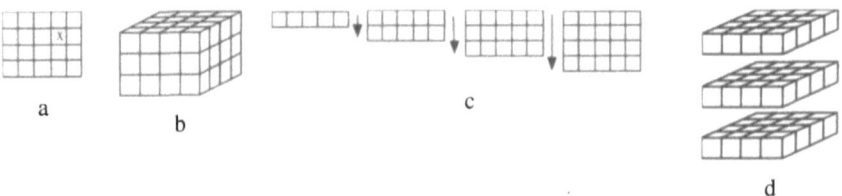

FIGURE 2   Arrays of squares and cubes (a & b); composite units(c & d).

dents can meaningfully enumerate arrays of squares and cubes only if, through the process of abstraction, they have developed properly structured mental models that enable them to correctly locate and organize the squares and cubes.

Two additional processes are required to construct such properly structured mental models. The *units-locating* process locates squares and cubes by coordinating their locations along the dimensions that frame an array. For instance, to understand the location of Square X in the array shown in Figure 1a, an individual must "see" the square in terms of a two-dimensional coordinate-like system—for example, it is in the fourth column and the second row, or it is the fourth unit to the right and the second unit down.

The *organizing-by-composites* process combines an array's basic spatial units (squares or cubes) into more complicated *composite units*[3] that can be repeated or iterated to generate the whole array. For instance, in a two-dimensional array, a student might mentally unite the squares in a row to form a spatial composite unit that can be iterated in the direction of a column to generate the array (Figure 1c). In a three-dimension array, the cubes in a horizontal layer can be grouped into a layer to form a spatial composite that can be iterated vertically to generate the array (Figure 1d).

Levels of Sophistication in Students' Structuring and Enumeration of Arrays

The four cognitive processes previously described are now used to describe levels of sophistication in students' understanding of area and volume measurement. The tasks used to reveal students' reasoning are highlighted with italics. Students' thinking about area is illustrated with second graders' work (Battista et al., 1998), and students' thinking about volume is illustrated with fifth graders' work (Battista & Clements, 1996; Battista, 1999).

To meaningfully and correctly enumerate arrays of squares or cubes, students must be capable of numerically operating on interiorized versions of appropriate mental models of these arrays. Thus, assessment tasks must provide information not only about the processes students use to enumerate arrays, but about the level of abstraction students seem capable of while operating on arrays. To reveal these levels of abstraction, two types of assessment tasks were used. In the first, all the spatial units that must be enumerated were perceptually available. That is, for arrays of squares, every square in the array is visible in the task diagram. In this case, students

---

[3] A composite unit is a cognitive entity (an abstraction) that results from mentally uniting and taking as one thing a collection of perceived or re-presented objects. Spatial composites are spatial entities formed from spatial objects. Numeric composites are numeric entities formed from items taken as countable units. Although research has identified levels of sophistication in students' development of numeric composites (Steffe, 1988, 1992), it is the development and use of spatial composites that is most relevant in analyzing students' thinking about area and volume. (For brevity, the term *spatial composite unit* is often shortened to *composite*.)

could correctly enumerate squares if they had a properly structured mental model of the array perceptually abstracted. For cube arrays, the situation is more complicated. In physical arrays of cubes, every cube is perceptually "available" only if, along at least one dimension, there are at most two layers; in pictorial representations, only single layer arrays have all the cubes perceptually available. Furthermore, although two-layer physical cube arrays seem to be easier for students to enumerate, not all cubes are visible at the same time, so, for most students, the situation is actually more like the situation in which not all units are perceptually available.

In the second type of task, not all the units are perceptually available. For arrays of squares, students operating on internalized mental models of the arrays need to draw squares to correctly enumerate them; students operating on interiorized mental models can correctly enumerate squares without drawing the missing squares. Except perhaps in the case of two-layer physical arrays, enumerating cube arrays—both physical and pictorial—always requires a properly structured mental model of the array abstracted at the interiorized level. (The interiorization requirement, along with the complexity added by a third dimension, are probably the two major factors contributing to the significantly increased difficulty students have with enumerating arrays of cubes versus arrays of squares.)

Level 1: Absence of units-locating and organizing-by-composites processes.

Students do not organize units into spatial composites, and, because they do not properly coordinate spatial information, their mental models of cube arrays are insufficient to locate all the units in arrays.

Area.   Task: The student is shown that a plastic inch square is the same size as one of the indicated squares on the 7 × 3 in. rectangle displayed in Figure 3a. The

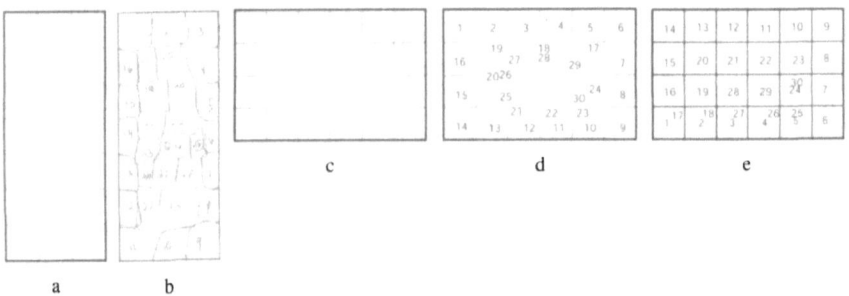

        a   b   c   d   e

FIGURE 3   Katy's work.

student is then asked to predict how many plastic squares it takes to completely cover the rectangle. Katy drew squares and counted 30 as shown in Figure 3b.

Task: The student is asked to predict, without drawing, how many squares it takes to cover the rectangle shown in Figure 3c. Katy pointed and counted as in Figure 3d, predicting 30. When checking her answer, she pointed to and counted plastic squares as shown in Figure 3e, getting 30. When she counted the squares again, first she got 24, then 27.

As can be seen from these examples, Katy had not yet mentally constructed a row-by-column structuring to organize and properly locate the squares. Instead, because Katy's mental model located squares along an almost random path, she got lost in her counting. Note that Katy seemed to be operating at the perceptual level; she could not keep track of her counting unless she perceptually marked squares (as in her drawing).

Volume.   Task: The student is asked how many cubes are needed to completely fill the box shown in Figure 4a. Bob counted the eight cubes shown in the box, then pointed to and counted six imagined cubes on the box's left side, four on the back, four on the bottom, and five on the top. His units-locating process was insufficient to create an accurate mental model of the cube array.

Task: The student is shown a picture of a box with the length, width, and height labeled and is told, "This box contains three cubes along the bottom, three up from here to here, and four from here to here [pointing appropriately at the box picture]. How many cubes does it take to completely fill the box. Draw what the cubes look like on the outside of the box." When Randa said she could not find the number of cubes needed to fill the box, she attempted to draw what the cubes looked like on the box. After about 10 to 15 min and many erasures, Randa's drawing looked like Figure 4b, showing a clear lack of coordination of spatial information. (e.g., the right side—inconsistently—shows five horizontal rows; the front shows only three.)

Task: The student is asked how many cubes it takes to make a cube building shown in a picture or made from interconnecting cubes (but they are not allowed to take physical buildings apart). The student is told that the building is completely

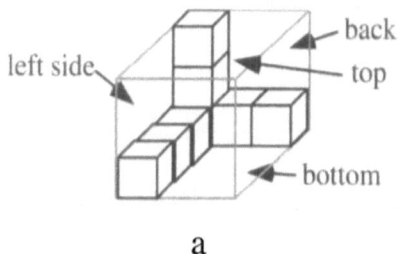

a

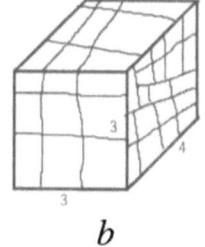

b

FIGURE 4   Level 1 volume work.

filled with cubes inside. The ubiquitous "double-counting" error on this task provides further evidence that insufficient coordination causes major difficulty in the units-locating process. For instance, when Jeff tried to enumerate the cubes needed to make the building shown in Figure 2b, he counted all cube faces that appeared on the six sides of the building, double-counting edge cubes and triple-counting corner cubes; he said there were two additional cubes in the interior. Because he could not properly coordinate what he saw on the different sides of the building, Jeff failed to realize when adjacent cube faces were part of the same cube.

Level 2: Beginning use of the units-locating and the organizing-by-composites processes

Students not only start to spatially structure arrays in terms of composite units, their emerging development of the units-locating process also produces mental models sufficient for them to recognize equivalent composites.

Area.   *Task: The student is shown that a plastic inch square is the same size as one of the suggested squares on the rectangle displayed in Figure 5. The student is asked to predict how many plastic squares it takes to completely cover the rectangle.*

Bill: First I count the bottom and there's six. [Moving his hands inward as shown in Figure 5] So the top and bottom would equal 12. And these two [pointing to the middle squares on the right and left sides] would be 14. [Using fingers to estimate where individual squares were located] I'd say maybe 12 in the middle; 12 + 12 = 24. So I'd say 24.

Bill was beginning to structure the array into composites (the top and bottom rows). Although he was unable to use the units-locating process to correctly locate interior squares, he did use it to see the numerical equivalence of his composites.

Volume.   For the building shown in Figure 2b—Fred counted 12 cubes on the front, then immediately said there must be 12 on the back; he counted 16 on the

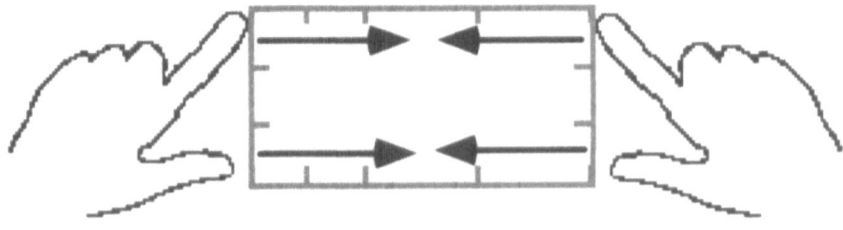

FIGURE 5   Bill's work.

top, and immediately said there must be 16 on the bottom; finally, he counted 12 cubes on the right side, then immediately said there must be 12 on the left side. In each case, after counting the cubes visible on one side of the building, he inferred the number of cubes in the opposite side, clear evidence that he was organizing cubes into composites and that he was using the units-locating process to relate these composites spatially and numerically.

Level 3: Units-locating process becomes sufficiently coordinated to recognize and eliminate double-counting errors.

A major breakthrough in thinking occurs when a student's units-locating process coordinates single-dimension views (e.g., top, side, front) into a mental model that is sufficient to recognize the same unit from different views. This refined mental model enables students to eliminate double-counting errors caused by insufficient coordination.

Area.   Task: The student is shown a 6 × 4 rectangle in which some of the squares are obscured (see Figure 6a). After showing that a plastic inch square is the same size as one of the suggested squares in the rectangle, the student is asked to predict how many plastic squares it takes to completely cover the rectangle. After correctly drawing the squares in the array, Bill counted the six squares in the left column, then immediately said there must be six in the right column. He counted four squares on the top, and then said there must be four on the bottom. He counted eight more squares in the middle, then added 6, 6, 4, 4, and 8 to get a total of 28. But when he explained his strategy to the interviewer, Bill changed his mind. He then said there were six squares in each of the left and right columns and counted two on the top, two on the bottom, and eight in the middle. Through an increase in coordination, Bill could simultaneously see a corner square as part of a row and as part of a column, enabling him to realize that he had initially double-counted such squares.

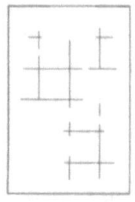

a

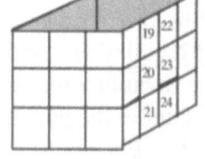

Counts 9 on the front, infers 9 on the back, making 18. Counts the 6 uncounted cubes on the right side, 19-24. Says there are 6 (uncounted) cubes on the left side.

b

FIGURE 6   Level 3 examples.

Volume. Task: *The student is asked to predict how many cubes of the size indicated by the unit squares on the side of the box it takes to completely fill the box shown in Figure 6b.* As shown in the figure, Juan coordinated spatial information sufficiently to avoid double-counting edge cubes. However, his coordination was still insufficient to build a mental model that properly located interior cubes.

Level 4: Use of organizing-by-composites process to structure an array with maximal composites, but insufficient coordination for iteration.

Students structure arrays in terms of maximal composites (rows or columns for area; layers for volume), producing more powerful and efficient mental models of arrays. For instance, a student might iterate a row of squares in the direction of a column to produce the whole array. But, due to insufficient coordination, they cannot precisely locate these composites.

Area. Task: *The student is shown that five plastic squares fit across the top of a rectangle and that seven fit down the middle (then the squares are removed).*

Joe: 5, 10, 15,..., 45 [motioning along estimated row positions inside the rectangle].
Tch: How did you get that?
Joe: I was trying to guess where the bottoms of the squares were.

Joe structured the array into row composites of 5. However, his coordination of rows and columns was insufficient to enable him to properly imagine the locations of the rows.

Volume. Task: *The student is shown a picture of a 5 × 3 × 4 cube array and is asked to build the bottom layer for the array, then to predict how many cubes it takes to make the whole array.* Randa built the 5 × 3 bottom layer and said that it contained 15 cubes. When asked how many cubes were in the entire building, she counted from the bottom up on the picture, but continued to count on the top, getting seven layers (see Figure 7). She gave an answer of 105 cubes. When the interviewer asked Randa to make the building with cubes, she built and stacked four layers, and was going to continue to build three more until the interviewer asked her to compare what she already built to the picture. Surprised, Randa concluded that the building was complete, then pointed to each layer saying, "15 here, 15 here, 15 here, 15 here; 154 = 60." Initially, Randa could not coordinate the horizontal layers with the third dimension that was the prism's height.

Although both Joe and Randa took major steps by structuring arrays in terms of maximal composite units, neither student had yet developed a sufficiently coordi-

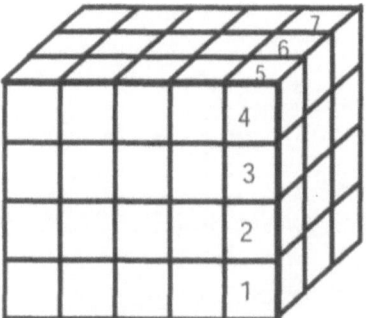

FIGURE 7   Randa's work.

nated and accurate mental model that allowed these composites to be properly iterated to form the whole array.

Level 5: Use of units-locating process sufficient to correctly locate all units, but less-than-maximal composites employed.

This is the first level in which the units-locating process is sufficient to create a mental model that correctly locates all squares/cubes in an array. However, although students sometimes get correct answers, because they inefficiently or inconsistently organize arrays into composites, they quite frequently lose their place in counting/adding and make enumeration errors. Furthermore, students' structuring and enumeration strategies are not generalizable and are inadequate for large arrays.

Area.   Task: The student is given the rectangle shown in Figure 8a and asked to predict how many unit squares are needed to cover it. Billie correctly employed the units-locating process (see Figure 8b). However, although she organized the array as composites of two squares (which was sufficient for correct enumeration), these composites were not the maximal ones needed to give the array its row-by-column structure.

Volume.   Task: The student is asked to predict the number of cubes it takes to make a 4 × 3 × 3 cube building shown in a picture (the student is told that the building is completely filled with cubes inside). As shown in Figure 8c, Mary counted the cubes visible on the front face (12), then counted those on the right side that had not already been counted (6). She then pointed to the remaining cubes on the top, and for each, counted cubes in columns of three: 1, 2, 3; 4, 5, 6; ... 16, 17, 18. She then added 18, 12, and 6. Like Billie, for area Mary correctly employed

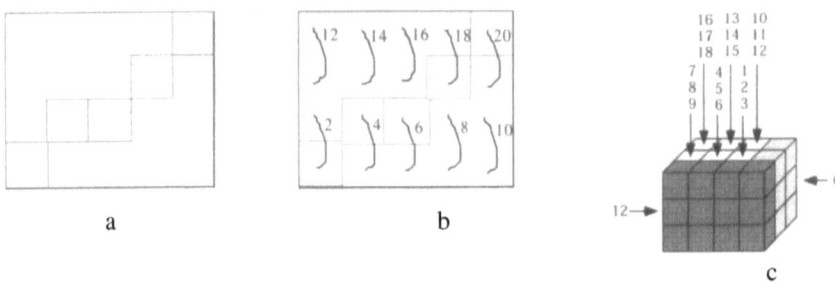

FIGURE 8   Level 5 examples.

the units-locating process. However, her structuring was overly complex and not easily generalizable.

Level 6: Complete development and coordination of both the units-locating and the organizing-by-composites processes.

Students' mental models fully incorporate a row-by-column or layer structuring so that they can accurately reflect on and enumerate an array without perceptual or concrete material for the individual units within composites.

Area.   Paul was shown that five plastic squares fit across the top of a rectangle and that seven fit down the middle (then the squares were removed).

Paul: [Counting and pointing across an imagined top row by ones] 5 across; 7 down. [Motioning across an imagined top 3 rows] 5, 10, 15. [Counting on seven fingers] 5, 10, 15, 20, 25, 30, 35; 35.

Int: How did you know to stop at 35?

Paul: There are only seven down that way [motioning vertically down the middle of the rectangle].

Volume.   For the cube array given in Figure 8c, Julie counted cubes in the top layer, 1 to 12, pointed to the middle layer and said 24, then pointed to the bottom layer and said 36. Her partner, Juanita, pointed to the right side of the array and said 9. She then counted four columns of three on the top and said, "So it's 9 times 4 equals 36."

Level 7. Students' spatial structuring and enumeration schemes become sufficiently abstract so that students can (a) understand the connection between numerical procedures and spatial structurings, and (b) generalize their reasoning to "packages."

(a) Students' spatial structuring and enumeration schemes reach a level of abstraction at which they can be reflected on and analyzed, thus enabling students to explicitly understand the connection between an enumeration strategy and the spatial structuring on which it is based. The following example illustrates students making a transition to such understanding.

Fifth-grader Bethany regularly determined the number of cubes in three-dimensional arrays using layers. She also had discovered that the number of cubes could be found by multiplying the length, width, and height. But as her class discussed this procedure, Bethany questioned its validity. She told her classmates that she was puzzled because "the corner cube gets counted once when you find the length, once for the width, and once for the height." Although almost every student in the class had discovered, and was routinely employing, a layer approach, not one of the students had an answer for Bethany's question. Even when the teacher posed Bethany's question in the context of area, the students had no answer.

The teacher took advantage of this teachable moment by having students work on the problem in pairs. When I asked Bethany and her partner how they were thinking about the problem, they said that they were "stuck." So I posed questions that I thought might help them clarify their thinking.

Me: [Arranges the 3 × 3 set of cubes that the students had been working with into three rows of three, and points successively to the cubes in one row] 1, 2, 3. What am I counting here?
Partner: Cubes.
Bethany: Yeah.
Me: [Pointing to the three rows] 1, 2, 3. What am I counting here?
Bethany: [Excitedly] Rows of cubes. You're not counting cubes this time. So, first, you count cubes, then you count rows.
Partner: So you're not really counting the cube twice. We got it!

Bethany's question posed a real conundrum for the students. They knew that multiplying the length times the width gave the number of cubes in a rectangular array. Almost all of the students justified this procedure by saying that they were multiplying the number of cubes in a row times the number of rows, thus satisfying the traditional criterion that they had learned the procedure "meaningfully." But initially, their understanding of this enumeration strategy did not clearly identify exactly what was being counted. To overcome their difficulty, Bethany and her partner not only had to properly structure the array, they also had to be able to ana-lyze their structuring so that they could properly conceptualize it.

Students at this level can also understand how to use linear measurements in enumeration procedures for arrays of squares and cubes. *Task: Give the student an unmarked box and explain that it is 5 cm wide, 8 cm long, and 3 cm high. Ask how many cubic centimeters it takes to completely fill the box.* To solve this problem with genuine understanding, students must recognize what these mea-surements imply about the locations of cubic centimeters, form a properly struc-tured mental model of a 5 × 8 × 3 array of cubes, and employ an appropriate enumeration strategy.

(b) Students' mental models incorporate row-by-column or layer structuring that is abstract and general enough to apply to situations in which the basic units are not cubes. The task shown in Figure 9 reveals students who have not attained this aspect of Level 7 (Battista, 2001a). At this level, students would not make the common error on this problem of multiplying 3 × 3 × 5, an error that indicates an insufficient connection between enumeration strategies and spatial structuring. In-stead, they would solve the problem by constructing a spatial structuring sufficient for correct enumeration.

CONCLUSION

To be consistent with scientifically sound views of mathematics learning, assess-ment must be properly linked to research on student learning and cognition. Be-cause this linkage is rarely made in traditional assessment paradigms (Masters & Mislevy, 1993), both instruction and assessment suffer. CBA takes a major step to-ward remedying this situation because it includes not only sound assessment tasks, but research-based conceptual frameworks for interpreting and understanding stu-

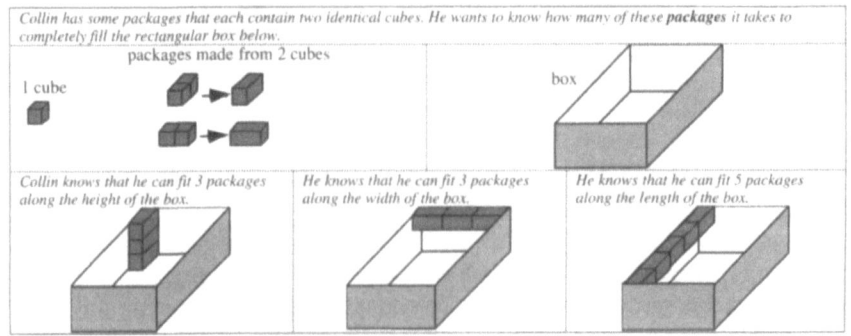

FIGURE 9   Two-cube problem.

dents' performance on these tasks. CBA helps us utilize the progress that has been achieved in modern research in mathematics education to significantly improve students' mathematics learning.

Indeed, once CBA tasks and levels-of-sophistication frameworks have been used to locate a student's position in learning a core idea, we, as teachers, know what cognitive processes and conceptualizations that student must acquire to make progress in constructing meaning and competence for the idea. Furthermore, we know what processes and conceptualizations are likely to occur next in students' development and can thus make reasonable conjectures about what instructional tasks might promote that development. Thus, by locating a student on a reasonably detailed map of the cognitive terrain required to construct understanding of a topic, CBA provides the knowledge needed for cognition-based instruction to successfully guide students in their construction of mathematical meaning.

Of course, the use of CBA could profit from additional research. First, levels of sophistication should be worked out for all major core topics. Second, the levels described in this article focused on the core measurement idea of enumerating arrays of squares and cubes. There are other core ideas related to area and volume measurement. Students must (a) develop understanding of how measures of non-rectangular figures can be determined, (b) be able to relate their understanding of area and volume measure to decomposing and recomposing figures, and (c) extend their thinking about area and volume measure to rational-number (vs. integral) measures. Additional research and levels are required to map out levels of sophistication in these areas, and to relate these levels to the levels presented in the current article. Furthermore, research is needed that relates the levels described for area and volume measurement to levels described for multiplicative thinking (see e.g., Steffe, 1988, 1992). Third, research is needed that elaborates currently proposed levels of sophistication—are there sublevels, what are the most common paths through the levels. Fourth, and finally, research should investigate the types of tasks and instructional guidance needed to help students move to higher levels of sophistication and should carefully trace instructed students' progress through the levels (e.g., Battista, 1999, traces students' learning progress in an instructional treatment that was based on early versions of the volume levels).

ACKNOWLEDGMENTS

This article is based on a presentation at the NCTM Research Presession, Las Vegas, April 2002. Support for this work was provided by Grant ESI–0099047 from the National Science Foundation. The opinions expressed, however, are the author's and do not necessarily reflect the views of that foundation.

REFERENCES

Battista, M. T. (1999). Fifth graders' enumeration of cubes in 3D arrays: Conceptual progress in an inquiry-based classroom. *Journal for Research in Mathematics Education, 30,* 417–48.

Battista, M. T. (2001a). How do children learn mathematics? Research and reform in mathematics education. In T. Loveless (Ed.), *The great curriculum debate: How should we teach reading and math?* (pp. 42–84). Washington, DC: Brookings Press.

Battista, M. T. (2001b). *The development of a cognition-based assessment system for core mathematics concepts in grades K–5.* Project funded by the National Science Foundation.

Battista, M. T. (2003). Understanding students' thinking about area and volume measurement. In D. H. Clements (Ed.), *2003 yearbook, learning and teaching measurement* (pp. 122–142). Reston, VA: National Council of Teachers of Mathematics.

Battista, M. T., & Clements, D. H. (1996). Students' understanding of three-dimensional rectangular arrays of cubes. *Journal for Research in Mathematics Education, 27,* 258–292.

Battista, M. T., Clements, D. H., Arnoff, J., Battista, K., & Borrow, C. V. A. (1998). Students' spatial structuring and enumeration of 2D arrays of squares. *Journal for Research in Mathematics Education, 29,* 503–532.

Battista, M. T., & Larson, C. N. (1994). The role of the *Journal for Research in Mathematics Education* in advancing the learning and teaching of elementary school mathematics. *Teaching Children Mathematics, 1,* 178–82.

Bransford, J. D., Brown, A. L., & Cocking, R. R. (1999). *How people learn: Brain, mind, experience, and school.* Washington, DC: National Research Council.

Carpenter, T. P., & Fennema, E. (1991). Research and cognitively guided instruction. In E. Fennema, T. P. Carpenter, & S. J. Lamon (Eds.), *Integrating research on teaching and learning mathematics* (pp. 1–16). Albany: State University of New York Press.

Carpenter, T. P., Fennema, E., Fuson, K., Fuson, K., Hiebert, J., Human, P., et al. (1999). Learning basic number concepts and skills as problem solving. In E. Fennema & T. A. Romberg (Eds.), *Mathematics classrooms that promote understanding* (pp. 45–62). Mahwah, NJ: Lawrence Erlbaum Associates, Inc.

Cobb, P., & Wheatley, G. (1988). Children's initial understanding of ten. *Focus on Learning Problems in Mathematics, 10*(3), 1–28.

Cobb, P., Wood, T., & Yackel, E. (1990). Classrooms as learning environments for teachers and researchers. In R. B. Davis, C. A. Maher, & N. Noddings (Eds.), *Constructivist views on the teaching and learning of mathematics. Journal for Research in Mathematics Education Monograph Number 4* (pp. 125–146). Reston, VA: National Council of Teachers of Mathematics.

De Corte, E., Greer, B., & Verschaffel, L. (1996). Mathematics teaching and learning. In D. C. Berliner & R. C. Calfee (Eds.), *Handbook of educational psychology* (pp. 491–549). New York: Simon & Schuster Macmillan.

Fennema, E., Carpenter, T. P., Franke, M. L., Levi, L., Jacobs, V. R., & Empson, S. B. (1996). A longitudinal study of learning to use children's thinking in mathematics instruction. *Journal for Research in Mathematics Education, 27,* 403–434.

Goldin, G. A. (1992). Toward an assessment framework for school mathematics. In R. Lesh & S. J. Lamon (Eds.), *Authentic assessment performance in school mathematics* (pp. 63–88). Washington, DC: AAAS Press.

Greeno, J. G., Collins, A. M., & Resnick, L. (1996). Cognition and learning. In D. C. Berliner & R. C. Calfee (Eds.), *Handbook of educational psychology* (pp. 15–46). New York: Simon & Schuster Macmillan.

Hiebert, J., & Carpenter, T. P. (1992). Learning and teaching with understanding. In D. A. Grouws (Ed.), *Handbook of research on mathematics teaching* (pp. 65–97). Reston, VA: National Council of Teachers of Mathematics/Macmillan.

Lester, F. K. (1994). Musing about mathematical problem-solving research: 1970–1994. *Journal for Research in Mathematics Education, 25,* 660–675.

Masters, G. N., & Mislevy, R. J. (1993). New views of student learning: Implications for educational measurement. In N. Frederiksen, R. J. Mislevy, & I. I. Bejar (Eds.), *Test theory for a new generation of tests* (pp. 219–242). Hillsdale, NJ: Lawrence Erlbaum Associates, Inc.

National Research Council. (1989). *Everybody counts.* Washington, DC: National Academy Press, 1989.

Romberg, T. A. (1992). Further thoughts on the standards: A reaction to Apple. *Journal for Research in Mathematics Education, 23,* 432–437.

Schoenfeld, A. C. (1994). What do we know about mathematics curricula. *Journal of Mathematical Behavior, 13,* 55–80.

Steffe, L. P. (1988). Children's construction of number sequences and multiplying schemes. In J. Hiebert & M. Behr (Eds.), *Number concepts and operations in the middle grades* (pp. 119–140). Reston, VA: National Council of Teachers of Mathematics.

Steffe, L. P. (1992). Schemes of action and operation involving composite units. *Learning and Individual Differences, 4,* 259–309.

Steffe, L. P., & Cobb, P. (1988). *Construction of arithmetical meanings and strategies.* New York: Springer-Verlag.

Steffe, L. P., & D'Ambrosia, B. (1995). Toward a working model of constructivist teaching: A reaction to Simon. *Journal for Research in Mathematics Education, 26,* 146–159.

Steffe, L. P., & Kieren, T. (1994). Radical constructivism and mathematics education. *Journal for Research in Mathematics Education, 25,* 711–733.

van Hiele, P. M. (1986). *Structure and insight.* Orlando, FL: Academic.

von Glasersfeld, E. (1995). *Radical constructivism: A way of knowing and learning.* Washington, DC: Falmer.

MATHEMATICAL THINKING AND LEARNING, 6(2), 205–226

Evolving Communities of Mind—In Which Development Involves Several Interacting and Simultaneously Developing Strands

Richard Lesh and Caroline Yoon
School of Education
Purdue University

If a curriculum developer's goal is to create a single linear sequence of tasks that lead to the development of some important mathematical concept, then some researchers have suggested that these sequences should follow progressions similar to stages of development that have been identified in Piaget-like research on the relevant concept(s). These research-based sequences are referred to as learning trajectories.

Other researchers emphasize that conceptual development can involve interactions among ideas expressed using a variety of representational media and can occur along a variety of "dimensions" such as concrete-abstract, simple-complex, or situated-decontextualized. Therefore, different paths can be appropriate for different students, and trying to funnel development along any single developmental path can be inappropriate for some students. These researchers often envision trajectories to be specific paths within a branching tree diagram that portrays the space of possibilities.

This article emphasizes a third type of situation called *model-eliciting activities* (Lesh, Hoover, Hole, Kelly, & Post, 2000). They are problem-solving situations in which goals include developing more powerful constructs or conceptual systems. Therefore, significant conceptual developments occur because students are challenged to repeatedly express, test, and revise their own current ways thinking—not because they were guided along a narrow conceptual path toward (idealized versions of) their teacher's ways of thinking. That is, development looks less like progress along a path, and it looks more like an inverted genetic inheritance tree in which great grandchildren trace their evolution from multiple lineages that develop simultaneously and interactively.

Requests for reprints should be sent to Richard Lesh, Purdue University School of Education, 100 N. University Street, West Lafayette, IN 47907–2067. E-mail: rlesh@purdue.edu

In effective mathematics learning environments, it often is important for teachers to play the role of guides who direct students along paths that lead efficiently toward the teacher's (or the textbook's) portrayal of the meaning of important ideas. In these situations, idea development often can be characterized as sequences of stages organized into learning trajectories that students typically pass through. This article goes beyond examining the different stages of ideas to also observe and describe the processes that contribute to idea development. In particular, we shift attention beyond situations in which students' thinking is guided by teachers and trajectories to focusing on activities in which students express, test, and revise their own ways of thinking.

We begin by critically examining two common metaphors for describing learning trajectories of idea development; ladder-like linear sequences and branching tree diagrams. Next, we present results from situations in which significant conceptual changes can be seen to occur during a single 60- to 90-min problem-solving session, but in which progress is not characterized by movement along a single path. Results show that solution processes usually draw on understandings and abilities associated with a variety of textbook topic areas, and they typically involve sequences of design cycles in which relevant ways of thinking are gradually sorted out, integrated, related, refined, and revised or rejected to fit the needs of specific situations and purposes. Finally, we introduce the language of evolving communities of mind to emphasize the assumptions that (a) thinking is based on communities of constructs (and conceptual systems) that are all at intermediate stages of development, (b) development resembles the evolution of a community of living, interacting, and evolving biological systems (Dawkins, 1990), and (c) certain kinds of development are unlikely to occur in the absence of Darwinian processes such as diversity (Minsky, 1987), selection (i.e., survival of stable and productive systems), communication (i.e., spread throughout the community), and accumulation (i.e., preserving adaptations so they apply to new situations).

LEARNING TRAJECTORIES AS METAPHORS FOR IDEA DEVELOPMENT

What Do Ladder-like Sequences Suggest About Idea Development?

Following the lead of Jean Piaget (Piaget & Beth, 1966), mathematics educators have clarified stages of development in students' understanding of important mathematical constructs in topics ranging from early number and measurement concepts, to rational numbers and proportional reasoning, and to algebra, geometry, calculus and statistics. However, unlike Piaget, who focused on the natural development of children's general cognitive structures throughout the time span of childhood in daily life experiences, these researchers have focused on the induced development of specific ideas in short periods of time in classroom environments (Lesh & Carmona, 2003). Researchers usually structure these stages of knowledge development in ladder-like sequences, with each rung corresponding to a more sophisticated understanding of the construct that is concerned. Such sequences have been useful for curriculum and software developers who wish to design textbooks that will guide students toward deeper understandings of important ideas. But when this model is used to make inferences about how ideas develop, it carries some of the following problematic assumptions.

Development occurs along one dimension. Ladder-like sequences imply that development occurs linearly (i.e., along one dimension). However, research has provided overwhelming evidence that mathematical constructs (and conceptual systems) develop along a variety of dimensions. For example, Piaget emphasized development from concrete operational systems of thought toward formal operational systems (Piaget & Beth, 1966). Vygotsky (1978) emphasized the internalization of external cognitive functions. Bruner (1960) emphasized increasing representational fluency and the development of progressively more powerful representational media in which to express current ways of thinking. Other researchers have emphasized further dimensions of conceptual development such as concrete–abstract, specific–general, global–analytic, simple–complex, situated–decontextualized, and intuitive–formal.

Development occurs within neat topic boundaries. Ladder-like sequences tend to be used to describe deepening understandings of at most one important mathematical idea. This can imply that students' understandings of ideas develop one at a time, without transgressing outside arbitrarily assigned topic boundaries. This is often reflected in textbooks that partition lessons into tidy topic areas, in which the goal is for students to learn mathematics by becoming proficient with one idea before moving onto another. In contrast to this topic organization that suits textbooks, when we talk to people who are heavy users of mathematics in fields in engineering, management and others, what we hear most often is that nontrivial problems seldom can be solved by drawing on ideas from a single topic area (Lesh, Zawojewski, & Carmona, 2003). One reason why this is true is that in realistic situations, productive models and conceptual tools usually need to deal simultaneously with multiple issues such as quality and costs in complex systems.

What Do Branching Tree Diagrams Suggest About Idea Development?

Branching tree diagrams have proven to be useful when it is possible to adapt teaching and learning activities to fit to the emerging ways of thinking of different students or groups. Such instances include classroom discussions in which teachers use Socratic (guided questioning) techniques to guide students' thinking about a given problem, concept, or topic area.

Tree diagrams have most of the same advantages as ladder-like sequences. They too can be informed by research about typical developmental paths to deeper or higher-order understandings and abilities associated with powerful concepts in elementary mathematics. But, because they provide a variety of alternative paths to any given goal, they encourage teaching and learning activities to be modified to fit the ways of thinking of different students or learning communities. Tree diagrams also have the advantage that they can be designed to reflect social aspects of development in addition to cognitive aspects that are based on research about the thinking of isolated children. This is significant because student development typically occurs through participation in learning communities (or "communities of practice") in which the conceptual development of individuals is strongly influenced by shared constructs, conceptual tools, and social norms of the group (Wenger, 1998). But, it is one thing to recognize that the thinking of individual students is strongly influenced by the communities of practice in which the students function, and it is quite another to imagine that all of the students are of one mind—so that it becomes possible for teachers to think of themselves as teaching to "the class" (taken as a single entity). For example, when students work in three-person teams on the kind of model-eliciting activities that are described in the following discussion in this article, or when they make presentations about the results of their work in class discussions, it nearly always becomes clear that there are a variety of different levels and types of equally useful solutions to nearly any realistically complex problem.

• Even though each three-person team of students can develop a shared, sharable, and reuseable way of thinking that represents the consensus perspective that is developed by the group, this collective thinking of the group-as-a-whole generally is not shared equally by all members of the group. In fact, at the same time that individual students support the consensus opinions developed by their group, it often is the case that they also continue to develop ideas, tools, and perspectives that are not shared by the group-as-a-whole.

• Different groups often develop significantly different consensus perspectives that do not necessarily coincide with the consensus perspective that develops for the classroom community-as-a-whole.

In mathematics classrooms, even the most insightful dialogs between the teacher and the class-as-a-whole can be quite different than simultaneous interactions that occur between the teacher and individual students. Furthermore, any time that one perspective is treated as being more "politically correct" than others—because it is favored by the teacher or by the community even though it might not be the most useful response based on "design specs" that were given in the statement of problems or tasks—there is a danger that some students' ways of thinking will be rejected consistently by the group (or the teacher) even though experts watching videotapes of the sessions could consider their contributions to be at least as valuable as those of their peers or patrons. Negative effects of teacher-imposed (or community-imposed) political correctness are being seen consistently in Purdue's *Gender Equity in Engineering Project* (Ashmann, Zawojewski, & Bowman, 2003), which focuses on students' performance in model-eliciting activities similar to those that are described in this article. In fact, the design principles for creating model-eliciting activities explicitly emphasize that evaluating students' solutions based on criteria different than those specified in the statement of tasks is one of the primary derailers of outstanding performance on such tasks.

Do Ideas Really Develop Along a Trajectory?

Many of the previously stated problems about the development of ideas stem from the way that trajectories identify the level of students' functioning abilities as stable across contexts. According to this perspective, a student once identified as a stage N thinker is expected to function at that stage across all other tasks that are characterized by the same underlying construct/conceptual system. Unfortunately, this assumption ignores an enormous research literature on task variables that influence task difficulty. Piaget (Piaget & Beth, 1966) used the term *decalage* to refer to these phenomena of variability across structurally similar tasks, and Vygotsky (1978) relied on it to account for his *zones of proximal development*. In general, research literature in mathematics education is filled with examples showing how the difficulty of a task often varies considerably if slight changes are made in a variety of task variables. For this reason, we find it useful to distinguish between research that results in generalizations about children and research that results in task-specific generalizations about ideas (Lesh, 1985). The former type of research results in claims such as "This is a stage N child (across all structurally similar tasks)." Whereas, the latter type of research results in claims such as "This level of understanding can be expected to follow and precede these levels of understanding on a specific task."

A MODELS AND MODELING PERSPECTIVE

To challenge the assumption that ideas develop along a trajectory, we examine a situation that is not artificially engineered to develop students' understandings along a single narrow conceptual path. That is, we need a situation in which (a) ideas are developing (b) without teacher/textbook guidance (c) in an observable way. Such situations are not easy to come by under the current climate of learning trajectories. However, these are exactly the kinds of situations that are consistently obtained using activities designed from the theoretical perspective we take: models and modeling (see Lesh & Doerr, 2003, for a more in-depth treatise of the perspective). The transcript that is presented in this article is an example of a response to one such activity, called "The Quilt Problem."[1] In this section, we explain how these activities reliably elicit observable idea development without direct teacher or textbook guidance.

What Kinds of Idea Development are Elicited by These Activities?

A models and modeling perspective assumes that mathematical thinking is about interpreting situations mathematically at least as much as it is about processing information that already exists in relevant mathematical forms. Models are the "things" that mathematicians use to interpret situations mathematically by quantifying, dimensionalizing, coordinatizing, systematizing, or (in general) mathematizing objects, relations, operations, transformations, patterns, regularities, or other systemic characteristics of learning or problem-solving situations. Students not only need to use existing constructs and conceptual systems, but they also often need to modify or extend them by integrating, differentiating, revising, or reorganizing their initial mathematical interpretations. That is, they need to become engaged in model development. Consequently, we refer to the activities that are designed to ensure this as *model-eliciting activities*.

How Do Model-Eliciting Activities Ensure that Significant Idea Development Occurs Without Direct Teacher Textbook Guidance?

Rather than having teachers direct the path of learning, these activities confront students with a mathematically rich yet realistic and meaningful situation in which students recognize the need to express, test, and revise their own initial ways of thinking into powerful constructs and conceptual systems. Furthermore, the activities are inbuilt with ways for students to realistically assess the quality of their own ways of thinking, without predetermining what their final solution should look like (Lesh et al., 2000). For example, in the quilt problem shown in the following, students know that they have finished when their quilt templates produce a quilt built to scale, in which all the pieces fit together, but they are not told what form their solution should take. Consequently, students working on these activities are able to develop their ways of thinking into powerful constructs and conceptual systems without being told which path to take.

---

[1]The Quilt Problem that was used in the study reported in this article was developed as part of a project called Packets, which was supported by a grant from the National Science Foundation (Katims & Lesh, 1994).

How Do Model-Eliciting Activities Ensure that Idea Development Can Be Observed?

Two additional design factors enable outside observers, like researchers or teachers, to observe the process of idea development as it occurs. First, the activities provide realistic conditions under which students are required to express and document their thinking. Instead of statements asking students to "explain your solution" that are often added onto many word problems as an afterthought, these activities provide a realistic "client" who asks for a general solution to a specific problem, in which the process is the product. For example, in the quilt problem, students are asked to describe how to make the quilt pieces templates not only for the picture quilt, but also for quilts of any design. Therefore, their final answer is a documentation of their problem solving process. Students are also required to work in teams of three, which increases the amount of internal thinking that is expressed externally, and this can be captured by videotape transcriptions. Second, these activities are designed so that they can be completed in a short space of time—typically between 60 to 90 min. This gives a remarkable advantage over much other research that traces incremental development in children's ideas over long periods of time. That is, the conditions in the activity act like a hothouse for growing ideas, in which the growth occurs in a short enough space of time that the whole process is realistically observable. Unlike Piaget, who could only identify the different stages of development of conceptual structures in different children, model-eliciting activities allow us to view the process of development that results in such stages.

An Example of a Typical Model-Eliciting Activity: The Quilt Problem

The Quilt Problem is described in greater detail in Lesh and Harel (2003). The activity begins with a newspaper article about a local quilting club. The article explains how quilters make templates for pieces like the diamond shape shown in Figure 1. Then, students are required to work in teams of three to write a letter that (a) describes procedures for making template pieces that are exactly the right size and shape for any quilt whose photograph they might find and (b) demonstrate how to follow the procedures by making templates for each of the pieces of the quilt that is shown here (the picture measures 4 5/8 in. by 5 11/16 in.) for a quilt sized 78 by 94 in.

METHODOLOGY

The data presented here were taken from a class of seventh grade remedial students at a large Midwestern inner-city middle school working on the Quilt Problem. The school was ethnically diverse, and nearly all of the students qualified for free lunch. The teacher in charge of the class was experienced with implementing

FIGURE 1 An Example Photograph of a Quilt

model-eliciting activities and led the class through the newspaper article the day before they worked on the problem. Then, students worked on the problem in teams of three during a 90-min problem-solving session. The next day, the teams presented their solutions in another 90-min class session, and the teacher facilitated a class discussion on the different solutions. We videotaped both of the sessions, took detailed field notes, and collected student work. In this article, we present a transcript of one group's (Ann, Barb, and Carla[2]) solution. A detailed transcript for one of the other teams is given in digital appendixes of Chapter 4 of *Beyond Constructivism: Models & Modeling Perspectives on Mathematics Problem Solving, Learning, and Teaching* (Lesh & Doerr, 2003).

Why Do We Choose to Study the Development of Ideas that Occur in Groups?

In the previous section we mentioned that having students solve model-eliciting activities in groups meant that students were more likely to externalize their thoughts, making it easier for researchers to study their thinking. However, we also choose to study groups for theoretical insights into problem solving—much the same way that others choose to use experts and novices, or gifted students and average ability students, as windows for studying problem solving. Although it certainly is true that three-person teams sometimes function in ways that are different than isolated individuals, the similarities appear to be more significant than differences for the purposes of this article. That is, in this article the claims that we are interested in making are about problem solving rather than problem solvers (and

---

[2]These names are pseudonyms.

about idea development more than child development). Therefore, it is somewhat irrelevant to us whether "the problem solver" is a group or an isolated individual. In the transcript that is described in this article, the problem solver is in fact a three-person team of average ability middle school students. Indeed, in model-eliciting activities, many similar phenomena tend to occur regardless of whether the problem solver that we examine is a single isolated individual, a three-person team, or a whole classroom learning community.

Why Do We Choose to Study the Development of Ideas in Only One Group?

The most important claims that are made in this article not only apply to most other groups of middle school students, but they also apply to readers of this article, which will be obvious to any reader that stops reading at this point and tries to generate a own solution to The Quilt Problem. You may want to work in a team of three so that you externalize your thinking. For example, your solutions almost certainly will go through several modeling cycles in which you think in fundamentally different ways about givens, goals, and possible solution steps. Furthermore, your early interpretations almost certainly will be rather barren and distorted compared with your later interpretations. That is, you'll gradually notice more relevant information, and you will also be likely to decide that some of your initial assumptions were incorrect or excessively limiting. Nonetheless, even if your solution to The Quilt Problem bears these similarities to the solution that is described in this section, it also is likely that your solution will draw on somewhat different ideas and procedures than those used by the middle school students.

RESULTS: ANNOTATED TRANSCRIPT OF A GROUP'S SOLUTION PROCESS

We organized the following transcript into sections corresponding to major shifts in students' ways of thinking about their solutions to the Quilt Problem.

Estimating Sizes of the Pieces Without Explicitly Measuring Any of the Pieces

Initially, the students did not really compare the sizes of the real quilt and picture. Instead, they treated the picture as if it was actually a real quilt. That is, they labeled the picture using measurements from the real quilt, and used these measurements to estimate the sizes of several pieces of the real quilt.

After the teacher read the problem to the class-as-a-whole, the three students began talking about the quilts that were shown in the problem statement and the accompanying newspaper article and about people they knew who had made similar quilts. Carla mentioned having seen an unfinished quilt that her neighbor was making. They spent 8 min talking about quilts while getting out their toolkit of calculators, rulers, and other relevant materials. Finally, the students settled down to think about details of the problem.

Barb: (Barb is looking carefully at the picture of the quilt.) The hard part's gonna be doin' this (Barb is pointing to one of the large squares on the picture of the quilt.) How big you think they are?

Ann: I do'know. How big's the whole thing? ... Says here ... What? (Ann is looking at the statement of the problem.)

Barb: It's ... um ... it's 78 by 94. That's it.

Ann: Oh yeh! So it's about, like, 78 this way (pointing to the short side of the quilt) and, like, 94 this way (pointing to the long side of the quilt). How many feet's that? ... Hmm. Let'me see.

Barb: Where's the calculator? I'll do it. ... (She picks up the calculator.) What's it again?

Ann: 'Bout 78. 78 ... uh ... in. ... That's, let's see, how many feet?

Barb: Here. I'll do it. 78. 7 ... 8. (She's punching the numbers into her calculator). ... Now what? Uh. ... We want feet. Right? ... So what?

Carla: 4 times 12 is, like, 48 (she calculates this in her head). ... So, 5 is ... uh ... 5 times 12 is 60. ... So, um, 6 is, um ... 6 is 72. What's 7? Um. 7 is 80 ... uh, 80-what? (She writes the numbers down on a piece of paper and calculates 7×12). 84. That's it. It's 84. (As she's calculating, Carla has been writing these results: $4 \times 12 = 48, 5 \times 12 = 60, 6 \times 12 = 72, 7 \times 12 = 84$ in a list on a piece of paper.) ... I can't make it again. Not another one. (By this she means that she'll go beyond 94 if she adds another 12 to 84.) ... Cause we can't go past 94. Ya, 94. ... So, there's, like, 10 in left. It's um ... It's ... 10 in. That's it. 10. ... So, it's 7 feet and 10 altogether. ... That's here (pointing to the long side of the quilt. She writes "7 feet 10" near the long side on the picture of the quilt.)

Barb: So, what's the other side? ... 5 feet maybe? Huh. 5. Uh. 5 ...

Carla: Let's see. 5 feet is ... um ... 5 times 12 is, what, uh, 60 maybe. Yeah, 60. So, what do we want? ... Um ... Let's see. ... 94. No. 78 it says. 74.... So, it's more.

Barb: What do ya mean? ... Oh, I get it.

Carla: 6 feet is 72 (in.). And we want what? Uh. What? Um. Here. 78. ... So that leaves what? Hmmm. ... 6 in. I guess. ... It's 6 feet and 6 in. on this side (pointing). ... (Carla writes "6 feet 6" on the short side of the picture of the quilt). ... So, it's more'n 7 feet this way, and more'n 6 feet this way. (long pause) ...Look. 1, 2, 3, 4, 5, 6, 7. ... It's about 7 on this side. (Carla is counting as if she's thinking of the quilt as being made up of a 6 × 7 array of squares as shown in Figure 2.) Now do it the other way. Let's see. 1, 2, 3, 4, and 5, and 6. ... And, it's about 6 on this side. So, these (pointing to the large squares) are about a foot big. See. ... They're 'bout a foot. ... These too. (She's pointing to the trim on the sides of the quilt.)

Barb: Ya, but these (the width of the trim) aren't as big as these (the width of the squares). ... That's not gonna work.

Carla: Uh huh. But, it's close. (Carla is still attracted by the idea that, according to her way of thinking, the quilt is (a) about 6 feet × 7 feet, and (b) about 6 squares × 7 squares, in spite of the fact that both of the preceding statements are not quite true.)

Ann: Where'd you count these ... these strips? (Ann is pointing to the dark strips that surround each of the large squares in the quilt.) ... If you count 'em, it's 'bout like the squares. (Ann is pointing out that the width of the trim plus the width of one of the dark strips is close to the width of one of the squares.)

Barb: Yep, you got somethin' there. ... But, it's gotta be exact. So the pieces fit together.

Ann, Barb, and Carla spent approximately 10 min more estimating and adjusting their guesses about the sizes of the large squares, the dark strips, and the trim. As they continued to use these guess-and-test methods, they became increasingly aware that the sizes of their guesses were not "adding up." For example, when they guessed that the large squares were 1 foot, and the dark strips 3 in., they found that this left 16 in. altogether for the trim on the length of the quilt, but only 15 in. for the trim on the width of the quilt. But they assumed that a "correct" method should result in a trim that would be exactly the same size on all four sides of the quilt. Be-

FIGURE 2   Quilt-as-a-Whole.

cause none of their "reasonable guesses" were yielding equally sized trims, the students decided that "something might be wrong" with their basic approach.

Comparing the Size of the Picture of the Quilt to the Size of Real Quilt by Determining a "Scaling-up" Factor

Next, the group recognized that The Quilt Problem was about "scaling-up"—like two of the other model-eliciting activities that they had worked on earlier in the semester (Lesh & Harel, 2003). They began to explicitly compare the sizes of the two quilts (i.e. the picture and the real quilt), and they began to look for "what to multiply by" to scale-up from one to the other.

Carla: Look. This is kinda like those other problems we did. Remember ... that big foot problem. Ya. That was neat. ... 'Member how we did it? ... All we gotta do is figure out what to multiply by.

Ann: So this picture is ... How big is it? ... Where's the ruler?

Barb: Here. I'll do it. ... It's ... um ... 4 in. and 1, 2, 3, ... , 9, 10. ... It's 4 × 10. (Barb is measuring the picture with a ruler, and she is pointing to the 4 5/8-in. side of the quilt.) ... 4 in. ... and 10 of these little marks.

Ann: How big are they? (Ann is asking what to call the markings on the ruler for 1/16 of an in.)

Carla: Tenths. ... I think they're tenths.

Barb: Tenths.

Carla: Okay. Tenths. So, what is it?

Ann: (Ann is ready to record the measures on the picture of the quilt.) Four and what?

Carla: 10.

Ann: 10/10. ... (Ann is writing "10" after the "4" on the picture. Then there is a long pause while she looks at it.) ... No way!! That's not it? Ten'd be a whole in. ... It's only 'bout half. A little more. ... (long pause) ... Look. How many of these little marks are in here (indicating a 1-in. gap on the ruler). 1, 2, 3, 4, ... , 16. There are 16 of 'em. ... So what are they? ... Oh God. Sixteenths, I guess. Is that it?

Carla: Yep, that's it. ... This side is 4 in. and 10 of these 16 things. Sixteenths of an inch. What're we gonna do with that?

Ann: How big is it really? (Ann is asking how big the comparable side of the real quilt should be.) ... Here. We wrote it. It's 6 feet 6. ... That's 'bout 6½ feet.

Barb: (Picking up the calculator) How you punch this in? (Barb is pointing to where Ann had written "4 in. and 10/16.")

Ann: Geez. I do'know. ... Oh God. ... It's about 4½. (Ann is holding a ruler beside the picture of the quilt.)

Carla: We can't do 4½. It's gotta be exact ... or the quilt won't work.

Ann, Barb, and Carla spend approximately 8 min more arguing about how to "punch 4 10/16 into a calculator." They also go back to remeasure the sides of the picture several times because they keep thinking that they must have measured something wrong. But, they keep getting the same results. They did not make a mistake measuring. They just did not know what to do with lengths like 4 10/16 in.

Explicitly Measuring the Parts of the Picture of the Quilt, Using a Ruler

Next, Ann (and later Barb and Carla) temporarily ceased worrying about comparing (or scaling-up from) the picture to the quilt and returned to comparing the sizes of the pieces within the picture of the quilt. This time, however, they used a ruler to measure the pieces, instead of making direct comparisons.

Their decision to use a ruler contributed to some confusion that ultimately led them to change their approach again. When Ann measured the picture of the quilt using 1/16-in. units, she got 74 × 91 as the size of the picture. However, when Ann repeatedly measured the separate pieces of the quilt, and then added their lengths, she kept getting results like 72 (1/16-in. units). This mismatch occurred because there was no way to use the ruler to make measurements of small pieces that would be sufficiently precise.

While Barb and Carla continue trying to figure out "what to multiply by" to scale-up from the picture of the quilt to the real quilt, Ann starts to use the ruler to measure the smaller pieces of the quilt. She has written 2 near several of the dark strips to indicate that she measured their widths to be "two little marks on the ruler" (i.e., 2/16 of an in.). Then, she also has written 12 by several of the large squares to indicate that she has measured their widths to be "12 little marks on the ruler" (i.e., 12/16 of an in.). Similarly, she has written 7 by the trim on the short side of the quilt, and she has written 9 by the trim on the long side of the quilt. So, even though she has not yet noticed what this means, she has just documented the fact that the trim at the top and bottom of the quilt is not the exactly same width as the trim on the sides. This shows that the group's earlier assumption—that the size of the trim is the same on all four sides of the quilt—is not true. Yet, this assumption was the main reason why the group thought that their first guess-and-test methods were not leading to correct results.

Ann: Look. I measured all these pieces.
Carla: Yeah. What're ya gonna do with 'em?
Ann: I'm gonna add 'em up to get the whole side.
Carla: How come. We know that. ... It's ... Le'me see. ... It's 78 in. That's 6 feet 6. (Carla doesn't seem to recognize that Ann is measuring the pic-

ture in 1/16-units rather than simply labeling the picture using measurements borrowed from the real quilt.

Ann: Well. I do'know. Let's see. She starts adding the widths along the short sides of the picture of the quilt. ... (mumbling) ... 7, 2, 12, 2, 12, 2, 12, 2, 12, 2, 7. ... That's, 9, 14, 14, 14, 14, and 7. ... So, 4 times 14 is ... it's 56. ... and 9 more is 65. And 7 more is 72. ... So, it's 72 wide. (She writes 72 to indicate the width of the short sides of the picture of the quilt.) ... Look guys, it's 72 wide. 72 of these little marks.

Barb: Ya, but so what? What're we gonna do with it.

Ann: How big did you want it? ... I mean the real quilt. How big do you want it? ... Le'me see. ... Look. We want 74 in. That's what I wrote here. Is that right? ... I wonder if I added those right. (Ann is reading the "74" from the place where she recorded the width of the picture in 1/16-in. units. But, she's reading this measure as if it was the size of the real quilt in 1-in. units.)

Long pause. ... For approximately the next 8 min, Ann, Barb, and Carla all used rulers to measure the sizes of the pieces within the picture of the quilt. After Ann demonstrated how to do it, all three students measured by counting the rulers' marks for sixteenths of an inch.

Ann: Look guys. Check this out. I don't see what I'm doin' wrong. I keep getting' 72 and I want 74. Somethin's wrong here somewhere. ... (Ann is not really differentiating measurements of the picture in sixteenths of an inch and measurements of the real quilt in in. But, she is noticing that the results yield similar numbers. So, she is making the incorrect assumption that the two measurements should in fact be the same, and she's also assuming that this mismatch must be due to some mistake that she's made in measuring or in calculating.)

Carla: Show me what you're doing. I'll watch.

Ann: See, I'm measuring each piece using these little marks (That is, Ann is using the 1/16-in. marks on the ruler). See, it's 7 here (pointing to the trim). ... And, 2 here (pointing to the dark strips). ... And it's 12 here (pointing to one of the large squares).

Carla: Uh huh. So how many is it altogether? ... Here. Le'me try it. (She takes the ruler from Ann and she counts 1/16-in. markings on the ruler to measure the short side of the picture of the quilt. She counts 74 sixteenth-in. markings). ... Okay, I got 74. What'd you get?

Ann: I got 72. ... Look. Do these. (Ann points to the measurements she has recorded for the small pieces, and she's asking Carla to check her measurements of each of the small pieces within the picture.)

Carla: Okay. I get 7 here (for the trim). ... That's what you got. ... (pause) ...
And 2 (for the dark strips). So that's okay. ... (pause) ... And 12 here
(the big squares). ... Yep, you measured 'em all right. ... You musta
added wrong.

Barb: I'll check 'em with the calculator. ... Read 'em out and I'll punch 'em in.

Ann: Okay. Ready. ... Here goes. ... 7, 2, 12, 2, 12, 2, 12, 2, 12, 2, 7. ...
What'd you get?

Barb: Got it. It's 72.

Carla: Le'me see that thing.

For approximately the next 4 min, Ann, Barb, and Carla go through the same
procedures again and again.

Carla: Yep. I got 72. ... Now le'me check those. (long pause) ... Yep, I agree
with those too.

Barb: So now measure the whole thing (pointing to the short sides of the picture of the quilt). What d'you get?

Carla: (Carla measures the sides of the picture of the quilt) I get 74 here, and
... let's see ... 91 here. ... What was I supposed to get?

Ann: These little pieces add up to 72.

Carla: What the heck is going on here? That can't happen. We gotta get the
same thing.

Ann: Oh God. What now?

For approximately the next 5 min, Ann, Barb, and Carla mix off-task talk
with comments about what they've done and what to do next.

Carla: You know, I don't think these dark things, these spacers, are really 2. I
think they're a little bit bigger than 2. We just can't tell using the ruler.

Barb: Yea, these are crappy rulers.

Carla: The marks just don't fit the picture. See.

Ann: Yea, I think you're right. ... So, what're we gonna do?

Explicitly Comparing Parts Within the Picture But Basing
Judgments on a "Standard Shape" from Which Other
Shapes Can Be Derived

Next, Barb and Carla took the large square as a "unit shape" from which other
shapes can be derived. That is, they would (a) scale-up a single shape (the large
square), and (b) use this single unit shape to generate all of the other shapes in the
quilt. For the first time, this way of thinking shifted the attention away from the
quilt-as-a-whole (where the big squares, the dark strips, and the trim are consid-

ered to be the most important pieces) and focused more on the smaller shapes within the big square (see Figure 3). Although this way of thinking is mathematically equivalent to scaling-up each of the pieces separately using a single scaling factor, this mathematical equivalence is not at all obvious to most students.

Carla: Okay, I think this big square is a foot. That's what we got every time. A foot. ... So, let's use it to make the other pieces.

Barb: Ya! ... I think we're onto something' now. ... I like it!

Carla: Okay. Let's try makin' those cut-outs. (When Carla refers to a "cut-out" she is thinking of the template pieces that the "client" in the problem statement asked them to make.) ... What do we make 'em with? Paper I guess. ... (long pause) ... Here's the paper. How big is it? (Carla is asking about the size of a standard 8.5 × 11 in. piece of paper.) ... (long pause) ... #!@!! It's not big enough. We gotta have bigger paper.

(Long pause. The students go to get some large pieces of poster-sized paper from their teacher)

Carla: Okay. We gotta draw a square that's one foot. ... (pause. Carla uses a ruler to draw a 1-foot square on a piece of poster-ized paper) ... There. ... Now we gotta use this to make the smaller pieces.

After Carla and Barb have drawn several large squares, they start to work independently—going in two somewhat different directions. Ann just watches both of them. Barb uses a ruler to draw lines inside the large square. The "most basic shapes" that she seems to be trying to draw are the smaller squares within the big square. But, the difficulty that arises for her is that guess-and-test procedures keep leading to squares that are either too large or too small to fit with the other pieces. Consequently, the diamonds keep ending up looking "not quite right" (like the one that is apparent in Figure 4c). Barb's reaction continues using guess-and-test pro-

FIGURE   3   Focusing   on   Big Squares

cedures repeatedly—even though this process is too time consuming to work well for guess-and-test methods.

Carla also is using a ruler to draw lines inside the large square. But, unlike Barb, the most basic shapes that she's trying to draw are the diamonds. The problem that arises for her is that her diamonds (or stars) leave spaces that are not square (see Figure 5c). Yet, Carla, like Barb, continues to use guess-and-test methods because it "seems to be getting somewhere" even though she virtually has to start over with each new attempt—a process that is too time consuming to provide effective guidance toward increasingly better results.

Making Pieces That Are the "Right Shape" Then "Stretching" Them to Fit With Other Pieces.

After watching Barb and Carla for more than 5 min, Ann looks in the group's tool box that contained rulers, compasses, tape, calculators, and other useful tools. But, unlike Barb and Carla, who started by making 1-foot squares, Ann goes somewhat off-task by using the geometry compass to make circles—and by making flowers like the one shown in Figure 6c.

Carla: (Carla is sounding tired and frustrated) What're you doin' Annie. Those're neat!

Barb: Ya, wha'ca doin'.

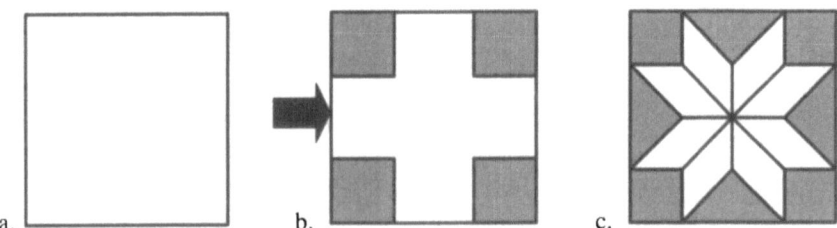

FIGURE 4 Barb treats Small Squares as the "Most Basic Unit Shape"

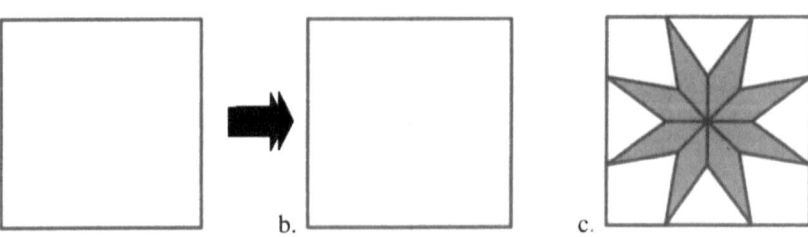

FIGURE 5 Carla treats the Diamonds as the "Basic Unit Shapes"

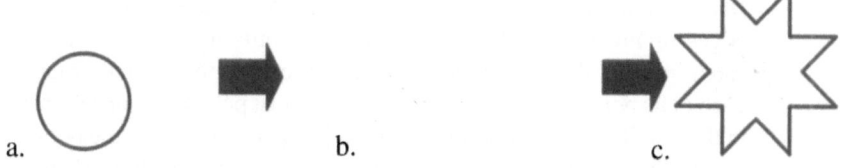

FIGURE 6   Ann Makes "Flowers" the "Right Shape" after Watching Barb and Carla.

Ann: I'm just watchin' you guys, and I like started making flowers—or stars, or whatever. ... Christmas, we made stuff like this for the tree like. ... Covered 'em with silver stuff—and punched a light in 'em. Looks good.

Barb: Ya, my mom likes stuff like that.

Carla: That one (Carla is pointing to one of Ann's stars) is just like what I need here (Carla is pointing to one of her squares that "didn't work".) ... How'd you do it anyway?

Ann: Here, I'll show ya. ... Pretty much like you did it. See here. (Ann points to two of Carla's squares that look similar to Figure 5a or 5b.) Just like you were do'in—only it's easier use'n a circle. (Ann is holding up a piece of paper similar to Figure 6a.)

Carla: I get it. ... Well, let's see. Let's just trace your star on my square. That'll do what we want. ... I think. ... Here, le'me try it. Gimme that one. (Carla is pointing to one of Ann's stars like the one in Figure 6c. Then, she puts it at the center of one of her own squares, as shown in Figure 7b.)

Barb: Nope. It's not quite big enough. Can you make it bigger?

Carla: I think maybe so. ... Hmm.

Barb: Ya. I got an idea.

Ann: Sure. It's easy. You can see right here exactly how it's gonna be. ... Look here. We just gotta stretch it out so the points come out here.

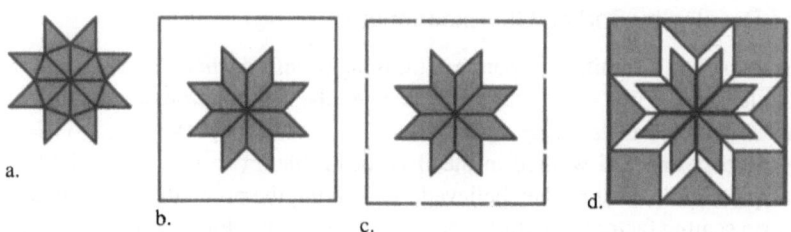

FIGURE 7   Carla Uses One of Ann's Stars in her Big Square Imagining Stretching to Fit

      (Ann is pointing to the positions where the "dots" are marked on the edges of Figure 7c. She appears to be identifying these points by "sighting out" along the dotted lines from the center of the star through each of its points.) ... I'll show ya. It's easy. Just put a good lookin' star in the middle of your square. Then, you just gotta make it so the points are this big. (Ann is pointing to the edge of the square.) But, it's just like this one. (Ann is pointing to the smaller star.) ... Fact. You can just trace it by lookin'. See. (Ann traces around a star like the one shown in Figure 7c. The result is a bigger star like the one in Figure 7d.) ... Now, just take the star away ... and that's it. (The result looks like Figure 3.)

Carla: I get it. It's like ... you just think like stretchin' it out to the edge. Makin' it bigger.

Ann: You got it. ... It's easy to make 'em bigger. Just get 'em right in the middle where they're suppose to be. Then, you can see how they should look bigger.

Barb: Come on you guys. We're 'bout outa time. ... And, we still gotta write our letter. ... What're we gonna say?

Carla: Main thing we gotta say is you gotta get one "key piece" that you use to make all the others. ... And, you gotta make it the right size. Like that big square. ... Then you just gotta make the other pieces using it. ... For little ones, it's easy to stretch 'em to fit. Like there. (Carla is pointing to a case like Figure 7d.)

In Ann, Barb, and Carla's final letter to the client, Barb emphasized Carla's points about (a) picking a key piece that can be scaled-up and (b) using this key piece to make other pieces. She also described Barb's idea of stretching difficult pieces to make them fit with other pieces.

THEMES ON THE PROCESS OF IDEA DEVELOPMENT
THAT EMERGE FROM THE TRANSCRIPT

Ideas Develop by Sorting Out Experiences

In the transcript, a significant change in thinking occurred when Carla realized that the Quilt Problem was about scaling up, just like the Bigfoot problem that they had completed a few weeks earlier. Having made this connection, they tried to use the same process that had worked in the Bigfoot Problem (Lesh & Harel, 2003) to scale up the quilt. That is, they believed that solving the problem merely required finding a scaling factor to multiply each of the pieces by. However, they soon realized that this would not work, because they did not have a sufficiently accurate way to measure each of the pieces. Even when Ann used an accurate ruler to measure the individual pieces, small errors were compounded when she multiplied them. Their final solution reinterpreted the scaling up idea by enlarging a single key piece—and then constructing other pieces using that piece as a unit (or building block). Qualitative changes in thinking can be seen by noticing changes in the mathematical objects and relations that the students attended to, and in their changing notions of the end-in-view (English & Lesh, 2003).

Ideas Develop by Adapting to Ends-in-View

Students initially interpreted the goal of the problem to be finding exact measures of individual pieces, and then scale these up to form the real quilt. Whereas later on, they reinterpreted their end-in-view as finding a way to recreate a scaled-up version of the quilt pattern from a key piece.

The students changed their perceptions of what kinds of thinking were useful when they compared their current solution to their current end-in-view. When their end-in-view was to scale up each individual piece of the quilt, they valued accuracy in measuring and calculations. But when their end-in-view changed to requiring them to recreate a scaled-up quilt from one key-piece, their contextualized knowledge became more important than abstract ways of thinking. Ann's contextualized Christmas star knowledge was a crucial part of their final solution. Instead of trying to develop their model toward dimensions that they may have considered as being universally good (e.g., abstract, complex, formalized), the students improved their model by trying to adapt to their current end-in-view.

Ideas Typically Trace Their Inheritance from Multiple Lineages

The final model that students came up with had traces of many different topic areas. It involved invariant properties of geometric shapes under transformation, proportions, measurement, calculations. Yet, final understanding was a unique model that did not fit neatly within any single one of these topic areas.

This can be visualized as inverted genetic inheritance trees in which great-grandchildren can trace their evolution from multiple lineages. Furthermore, these lineages, or mathematical topics, simultaneously and interactively developed into the students' final model. It is unlikely that the students would have had as rich an understanding had they been led along one path in, say, proportions and then told to make connections to geometry.

As Figure 8 suggests, ladder-like sequences can be thought of as specific options within a branching tree diagram, and similarly branching tree diagrams also can be thought of as the lineage from a specific ancestor within a genetic inheritance tree. According to this perspective genetic inheritance trees explain why concepts might appear to develop along linear paths or branching trees if attention fo-

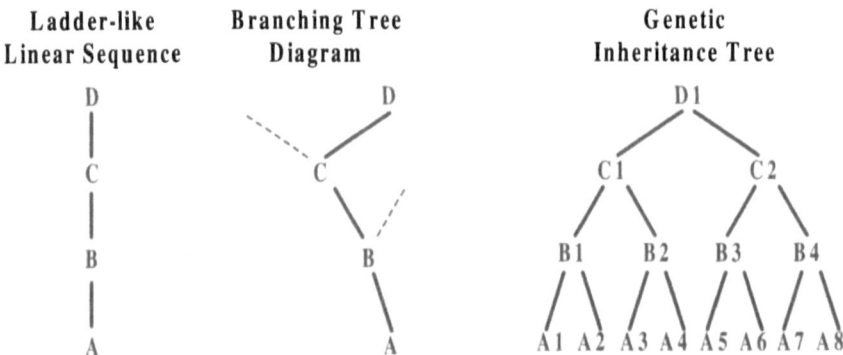

FIGURE 8   Ladder-like Linear Sequences, Branching Tree Diagrams and Genetic Inheritance Trees

cuses on narrow components of development. But, it also is clear that the reverse statement is not true. That is, linear sequences ignore options that branching trees make explicit, and branching trees ignore developmental roots that genetic inheritance trees make explicit. Furthermore, genetic inheritance trees suggest that the driving forces behind development are significantly different than those that are suggested by both branching trees and linear sequences—in which guidance by teachers (or learning communities) tend to be treated as the only way that significant forms of development are likely to occur. That is, genetic inheritance trees emphasize the integration and differentiation of diverse ways of thinking, and they suggest why guidance along a narrow conceptual path is not likely to be effective if too many genetic roots have been ignored. In other words, N-dimensional "regions of proximal development" are apparent which are similar to the one-dimensional zones of proximal development described by Vygotsky (1978).

TOWARD A NEW METAPHOR OF IDEA DEVELOPMENT: EVOLVING COMMUNITIES OF MIND

Far from being inert objects picked up by sufficiently competent students, the transcript suggests that ideas can be seen to function like biological organisms which develop by sorting out experiences, adapting to new environments, and inheriting mathematical traits from multiple lineages. We describe this development as *evolving communities of mind*. Under this metaphor, the classroom is characterized as an ecological system, in which communities of constructs and conceptual systems are struggling for survival. And like other types of complex, dynamic, interacting, and continually adapting systems, the development of ideas occurs in the presence of diversity, selection, reproduction, and communication. To take advantage of this, activities need to be designed which ensure that a diversity of ideas are expressed and that productive ideas are selected and preserved in conceptual tools and other classroom artifacts. In such situations, a teacher's role is not to prematurely guide students down specific or possible paths of learning, but to ensure that ideas are communicated, or diffused, within the classroom ecosystem.

REFERENCES

Ashmann, S., Zawojewski, J., & Bowman, K. (2003). *Integrated mathematics and science teacher education courses: A modeling perspective.* Manuscript submitted for publication.

Bruner, J. (1960). *The process of education.* New York: Vantage.

English, L., & Lesh, R. (2003). End-in-view problems. In R. Lesh & H. M. Doerr (Eds.), *Beyond constructivism: Models and modeling perspectives on mathematics teaching, learning, and problem solving* (pp. 297–316). Mahwah, NJ: Lawrence Erlbaum Associates, Inc.

Dawkins, R. (1990). *The selfish gene.* New York: Oxford University Press.

Katims, N., & Lesh, R. (1994). *PACKETS: A guidebook for in-service mathematics teacher development.* Lexington, MA: D. C. Heath.

Lesh, R. (1985). Conceptual analysis of problem-solving performance. In E. A. Silver (Ed.), *Teaching and learning mathematical problem solving: Multiple research perspectives* (pp. 309–329). Hillsdale, NJ: Lawrence Erlbaum Associates, Inc.

Lesh, R., & Carmona, G. (2003). Piagetian conceptual systems and models for mathematizing everyday experiences. In R. Lesh & H. M. Doerr (Eds.), *Beyond constructivism: Models and modeling perspectives on mathematics teaching, learning, and problem solving* (pp. 205–222). Mahwah, NJ: Lawrence Erlbaum Associates, Inc.

Lesh, R., & Doerr, H. M. (Eds.) (2003). *Beyond constructivism: Models and modeling perspectives on mathematics teaching, learning, and problem solving.* Mahwah, NJ: Lawrence Erlbaum Associates, Inc.

Lesh, R., & Harel, G. (2003). Problem solving, modeling and local conceptual development [Monograph]. *International Journal for Mathematical Thinking and Learning.*

Lesh, R., Hoover, M., Hole, B., Kelly, A., & Post, T. (2000). Principles for developing thought-revealing activities. In A. Kelly & R. Lesh (Eds.), *The handbook of research design in mathematics and science education.* (pp. 591–645). Mahwah, NJ: Lawrence Erlbaum Associates, Inc.

Lesh, R., Zawojewski, J. S., & Carmona, G. (2003). What mathematical abilities are needed for success beyond school in a technology-based age of information? In R. Lesh & H. M. Doerr (Eds.), *Beyond constructivism: Models and modeling perspectives on mathematics teaching, learning, and problem solving* (pp. 205–222). Mahwah, NJ: Lawrence Erlbaum Associates, Inc.

Minsky, M. (1987). *The society of mind.* New York: Simon & Schuster.

Piaget, J., & Beth, E. (1966). *Mathematical epistemology and psychology.* Dordrecht, The Netherlands: Reidel.

Vygotsky, L. S. (1978) *Mind in society: The development of higher psychological processes.* Cambridge, MA: Harvard University Press.

Wenger, E. (1998). *Communities of practice: Learning, meaning, and identity.* Cambridge: Cambridge University Press.

MATHEMATICAL THINKING AND LEARNING, 6(2), 227–260

Comments on the Use of Learning Trajectories in Curriculum Development and Research

Arthur J. Baroody, Michael Cibulskis, Meng-lung Lai, and Xia Li

College of Education
University of Illinois at Urbana-Champaign

In this commentary, we first outline several frameworks for analyzing the articles in this issue. Next, we discuss Clements and Sarama's overview and the issue hypothetical learning trajectories (HLTs) in general. We then analyze each of the other contributions. We conclude our commentary by offering a vision of HLTs that includes a key role for "big ideas."

FRAMEWORKS

We base our analysis of the articles in this issue, in part, on three frameworks. One is different philosophies of knowledge and authority (see Table 1), which provide the rationales for different approaches to mathematics instruction, the second framework summarized in Table 2. The third framework is a summary of Dewey's (1963) *Experience and Education* (see Table 3).

Philosophical Perspectives and Approaches to Mathematics Instruction

At one end of the direct-to-indirect instruction continuum is the traditional skills approach. Consistent with a dualistic philosophy, a teacher in this approach serves as the authoritative source of knowledge and uses direct instruction and practice to impart the correct procedure. The aim of such an approach is the mastery of basic skills (procedural content).

Requests for reprints should be sent to Arthur J. Baroody, College of Education, University of Illinois at Urbana-Champaign, 1310 S. Sixth Street, Champaign, IL 61820. E-mail: art-baroody@uiuc.edu

TABLE 1
Four Philosophical Views of the Nature of Knowledge and Authority

Philosophical Basis	Nature of Knowledge	View of Authority
Dualism	Right or wrong with no shades of gray: There is one correct procedure or answer.	Absolute external authority: As the expert, the teacher is the judge of correctness. Procedures or answers that differ from those advocated by the teacher are wrong and not tolerated. Teacher provides definitive feedback (e.g., praise for the correct answer).
Pluralism	Continuum from right to wrong: There is a choice of possible but not equally valid procedures or answers. Objectively, there is one best possibility.	Tolerant external authority: Teacher accepts diverse procedures and answers, but strives for perfection—namely, learning of the best procedure or answer. Teacher provides feedback (e.g., praises all ideas, particularly the conventional one).
Instrumentalism	Many right choices: There is a choice of possible procedures or answers and often many are good.	Open internal authority: Teacher or student remains committed to a method or viewpoint as long as it is effective. Teacher responds to incorrect procedures or answers by posing a question, problems, or task that prompts student reflection.
Extreme relativism	No right or wrong: There are many possible, equally valid possibilities.	No external authority: Teacher and each student define his or her own truth. Children evaluate their own conclusions.

At the other end of the continuum is the laissez-faire problem-solving approach. This is a process-oriented approach in that the aim is to develop mathematical thinking; learning content is secondary and incidental. As its underlying philosophy is extreme relativism—a teacher neither imposes solution procedures nor provides feedback on the correctness of solutions.

The first of two intermediate approaches is the conceptual approach, the aim of which is mastery of basic skills with understanding (i.e., its focus is conceptual and procedural content). A teacher can use, for example, highly structured guided discovery learning to lead students in a predetermined direction. Consistent with a pluralistic philosophy, teachers can tolerate or even encourage alternative procedures, but they ultimately ensure the standard procedure is adopted.

The second intermediate approach is the investigative approach. As a blend of the conceptual and problem-solving approaches, its aims are mastery of basic skills, conceptual learning, and mathematical thinking (i.e., its focus is conceptual and procedural content and process). The investigative approach, then, is characterized by both meaningful and inquiry-based instruction and by purposeful learning and practice. That is, a teacher uses worthwhile tasks to create a need to explore and use mathematics. As this approach is based on a philosophy of instrumentalism, teachers are

TABLE 2

Four Approaches to Mathematics Instruction (Baroody with Coslick, 1998)

Instructional Approach	Philosophical View	Teaching Style	Aim of Instruction
Skills approach	Dualism	Completely authoritarian and extremely teacher-centered: Direct instruction (teaching by imposition)	Foster routine expertise: The rote memorization of basic skills (arithmetic and geometric facts, definitions, rules, formulas, and procedures)
Conceptual approach	Pluralism	Semi-authoritarian and teacher centered: Direct and semi-direct instruction (teaching by "careful" imposition)	Foster adaptive expertise: The meaningful memorization of facts, rules, formulas, and procedures
Investigative approach	Instrumentalism	Semi-democratic and student-centered: Semi-indirect instruction (guided participatory democracy)	Foster all aspects of mathematical proficiency: Positive disposition (e.g., interest, confidence, and constructive beliefs about learning and using mathematics); Adaptive expertise (understanding of concepts and skills), and mathematical thinking (the capacity to conduct mathematical inquiry including problem solving and reasoning)
Problem-Solving approach	Extreme relativism	Completely democratic and extremely student-centered: Indirect instruction (teaching by negotiation)	Foster mathematical thinking: The ability to conduct mathematical inquiry

TABLE 3
A Summary of Dewey's (1963) Criteria for Successful Reform

---

1. Educational reform cannot simply be a knee-jerk reaction to traditional instruction (a skills approach). That is, new teaching methods cannot be substituted for traditional methods merely because they are different from the latter. New teaching approaches, methods, or tools must have their own (theoretical, empirical, and practical) justification. The *Principles and Standards of School Mathematics (PSSM;* NCTM, 2000), particularly Chapters 1 ("A Vision for School Mathematics") and 2 ("Principles for School Mathematics")—along with previous NCTM (1989, 1991, 1995) standards documents—provides a well-articulated justification for current reform efforts.

2. Instruction cannot simply consist of a hodgepodge of activities without clear educational purposes. Teachers must strive to provide *educative experiences*, experiences that lead to learning or a basis for later learning, not *mis-educative experiences*, experiences for experience sake and that may actually impede development. This sentiment is reflected in the following statement in the *PSSM* (NCTM, 2000): "High-quality learning results from formal and informal experiences during the preschool years. 'Informal' does not mean unplanned or haphazard" (p. 75). It is further reflected in the "curriculum principle:" "A curriculum is more than a collection of activities: it must be coherent, focused on important mathematics, and well articulated across the grades" (NCTM, 2000, p. 14).

3. Educational activities should not be unplanned or overlook external factors such as how a teacher can guide learning. Educative experiences result "from an interaction of external factors, such as the nature of the subject matter and teaching practices, and internal factors, such as a child's [developmental readiness] and interests" (Baroody, 1987, p. 37). Teachers, then, must ensure the harmonious interaction of these factors (e.g., considering the developmental readiness when choosing topics, activities, and teaching methods). The importance of both external and internal factors is emphasized throughout the *PSSM* (NCTM, 2000). For instance, the following quotes are clear allusions to the latter factor: "Teachers of young students…need to be knowledgeable about the many ways students learn mathematics" (p. 75). "Teachers must recognize that young students can think in sophisticated ways" (p. 77).

---

concerned about students' understanding and promote the use of any relatively efficient and effective procedure as opposed to a predetermined or standard one.

Dewey's Views on Reform

In *Experience and Education,* Dewey (1963) outlined the lessons he learned from his teaching experiments with a progressive philosophy (akin to extreme relativism). These lessons are still valuable today in evaluating efforts to reform (see Table 3).

LEARNING TRAJECTORIES IN MATHEMATICS EDUCATION (CLEMENTS AND SARAMA, THIS ISSUE)

Efforts to define instructional or learning courses are not new, but recent efforts to detail HTLs represent a significant improvement in such efforts. More specifically, unlike previous efforts to define learning sequences, HLTs entail a combination of all of the following: (a) goals for meaningful student learning, (b) tasks geared to achieve this learning, and (c) hypotheses about the process of student learning (local learning theory).

Previous Efforts to Define Learning Sequences

Learning sequences or hierarchies have long been a concern of educators and psychologists interested in school learning. Thorndike (1922), for instance, concluded that much time and effort was wasted because instruction did not take into account the laws of learning (e.g., the "law of effect"). He argued that, by knowing what bonds (associations) facilitated later learning and which interfered with it, instruction and practice could be better organized to promote learning. His findings or recommendations could be viewed as a series of goals. His general learning principles, such as the law of effect, can be viewed as hypotheses about learning processes of any given content. These principles, though, were global (as opposed to local) hypotheses and focused on simple associative, not meaningful, learning. Furthermore, Thorndike did not specify tasks to achieve such learning.

Similarly, the learning hierarchies developed or inspired by Gagné (e.g., Gagné & Briggs, 1974) essentially spell out a hierarchy of goals. These hierarchies were initially based on logical and empirical task analyses. The former entailed a logical consideration of what simple tasks were required to perform a complex task. Empirical task analysis entails collecting data to validate or adjust learning hierarchies. Gagné's learning hierarchies were used to create scope and sequence charts common to elementary textbooks. This and associative-learning theory suggested a sequence of direct instruction and practice exercises. In effect, like Thorndike's efforts (1922), Gagné's learning hierarchies focused on nonmeaningful learning and basically represented an improved skills approach.

More compatible with an emerging cognitive psychology at the time, Bruner (1966) proposed a discovery learning approach in which students would be encouraged to discover principles for themselves. He specified that a theory of instruction must address the following three issues:

1. predisposition to learn (i.e., experiences and contexts that will tend to make the learner willing and able to learn);
2. structure of knowledge (i.e., ways in which a body of knowledge should be structured so that the learner can readily grasp it); and
3. sequence (i.e., the most effective sequences in which to present the materials).

In some respects, then, Bruner's views seem most compatible with a conceptual approach although he was interested in helping students construct more powerful representations of knowledge and thinking.

In the last three decades of the twentieth century, as cognitive theory came into prominence, a new approach evolved. Cognitive or rational task analysis involved "analyzing tasks (goals) logically and intuitively, using what psychologists and educators already know" (Resnick & Ford, 1981, p. 58). More specifically, these efforts, including computer models of problem solving and development, built on information-processing theories and provided an important step toward supplying hypotheses about the processes of student learning of specific content.

Even so, efforts based on information-processing theory often did not focus on the construction of meaningful concepts or were based on the assumption that conceptual understanding can be imposed on children by, for example, showing students a manipulative-based concrete model and having them imitate it. In effect, these efforts, at best, embodied a pluralistic philosophy and exemplified a conceptual approach. Although more effective than the skills approach, such an approach often produced disappointing results (see Baroody, 2003, for a more complete discussion of the relative strengths and weaknesses of the conceptual approach).

There have been a number of efforts to use a combination of theories, including information-processing theory and a constructivist view, to prescribe developmental sequences (e.g., Griffin, Case, & Siegler, 1994). Baroody (1989a) used theory and research to layout a suggested instructional sequence of number and arithmetic concepts and skills, and Ginsburg and Baroody (2003) did the same in constructing a test of early mathematical ability. In these cases, specific instructional activities were proposed to foster the learning of each competence (or goal), and theory and research were used to provide guiding hypotheses about this learning.

Although the efforts based on cognitive theory (e.g., Baroody, 1989a; Resnick & Ford, 1981) approximated current HLT efforts, the developmental paths tended to be simple linear or ladder-like sequences. Moreover, the premise of instruction was closer to a conceptual approach than an investigative approach.

HLTs

Below we briefly discuss the three components of HLTs and how they represent an advance over previous efforts.

Goals.   In the first phase of HLT development, the initial trajectory or set of goals is developed using a variety of sources including, as Clements and Sarama noted, mathematical history and research on children's informal development research, which has expanded enormously over the last three decades. Another key source is constructivist theorizing and research, which provides a more detailed basis for specifying learning goals than earlier theories. As Battista suggested, yet another source is the mathematics viewed as socially important to learn and research on formal mathematical teaching and learning. In brief, current efforts draw on a wider and deeper range of sources for detailing goals than earlier efforts.

The second phase of many HLT development efforts is unlike nearly all previous efforts to define learning sequences. Educational experiments are used to evaluate the initial and subsequent HLTs and revise them accordingly. This systematic effort leads to the third phase, what Clements and Sarama called "a 'best-case' instructional sequence" but which might better be called "a relatively, potentially, or hypothetically 'best-case' instructional sequence." The latter is more in keeping with the view stated by these and other authors that learning trajectories are hypothetical and subject to constant revision.

The learning tasks.   Unlike previous efforts to specify learning sequences, many HLT efforts have as a goal the mutual development of theory and curriculum. At the same time theory is advanced, HLT researchers are using it to make practical applications (e.g., the systematic development, evaluation, and refinement of instructional activities, curriculum, and assessment tools). Another related key difference concerns ecological validity. In the past, research focused largely on individually administered tasks outside the classroom setting. HLT studies often entail evaluating a task in the context of ongoing instruction. Even when the focus is case studies, as reported in the Steffe article, a key aim is to find instructional implications.

Hypotheses about learning process.   Current HLT efforts are unlike many, or even most, previous efforts to define learning sequences in a number of important ways. (a) They are based on more extensive theory and research; (b) HLT researchers usually investigate development over time and developmental transitions; (c) HLT efforts build on a constructivist perspective and research to detail the evolution of meaningful learning; and (d) as was the case for task development and evaluation, because much of the HLT research is based on small group or classroom settings, the local learning theory developed has better ecological validity (i.e., it is more directly linked to what can be expected in real classroom instruction). Thus, HLT efforts provide a richer description of children's requisite knowledge, development, and difficulties than previous efforts to define learning sequences.

A Caution About HLTs

All research entails trade offs between internal and external validity. However valid the conclusions about a teaching experiment with individual children, small groups of children, or whole classes, ecological validity comes at a cost to generalizability. Clements and Sarama correctly pointed out, then, that an HLT cannot be considered the only path or even the best of many paths. In a real sense (and consistent with an instrumental philosophy), HLTs might better be called only a hypothesized possible learning trajectory, even when such a path or paths are confirmed with tens, hundreds or thousands of cases.

LOCAL INSTRUCTIONAL THEORIES AS A SUPPORT
FOR REFORM (GRAVEMEIJER, THIS ISSUE)

Gravemeijer (this issue) correctly noted constructivism is a catalyst of the current reform movement and an implication of this perspective is that instruction should capitalize on students' inventions (or reinventions). He further noted that, as building on students' invention and input seems incompatible with instructional planning, many interested in reform initially had little interest in learning hierarchies. This is because the designers of such hierarchies began their analysis with what experts know and worked backwards to detail each small step toward this ultimate goal (e.g., Gagné & Briggs, 1974). This is undoubtedly the case with logical and empirical task analyses.

Why Learning Sequences Were Rejected and Are Now
Again Embraced

However, why were rational task analyses based on cognitive theory and an effort to describe learning processes also shunned? Such an approach entails considering experts' knowledge, but it does so as a tool for considering what children must learn (the ultimate goal of instruction). In this sense, rational task analysis based on cognitive theory and current efforts to develop HLT based on a constructivist perspective are alike. That is, specifying increasingly more sophisticated ways of knowing and thinking and proposing a series of learning goals entails understanding the direction and ultimate goal of learning.

One difference, at least in a degree, between efforts based on rational task analyses and current HLT efforts is that those involved in the latter focus on how to promote developmental shifts (e.g., identifying and refining learning tasks). A second difference noted earlier is that hierarchies based on information-processing theory often do not adequately take into account or describe conceptual development, particularly the micro-conceptual development that is the bread and butter of classroom instruction. However, neither of these explanations explains why rational task analyses were initially forsaken (e.g., not even used as a starting point).

We suspect that radical constructivists or faith in this perspective heavily influenced the reform movement initially. The philosophical view underlying this view (extreme relativism) was at odds with that underlying logical/empirical task analyses (dualism) or even rational task analyses (dualism, pluralism, or some mix of the two, depending on the particular scholar). Put differently, initially the reform movement was influenced by those who favored a laissez-faire problem-solving approach. With its emphasis on invention of procedures over memorizing algorithms and justifying answers over efficient production of answers, for example, the reform movement in general was characterized, fairly or not, as a groundless and even dangerous reaction to traditional instruction (e.g., Ginsburg, Klein, & Starkey 1998; cf. Point 1 in Table 3).

As a result of criticism from reactionaries and the reflection of reformers themselves (e.g., Cobb, Wood, & Yackel, 1991), moderate constructivism became more influential in the reform movement. From this perspective, instruction needs to be a combination of the best aspects of a meaningful (content-oriented) conceptual approach and an inquiry-based (process-oriented) problem-solving approach (i.e., needs to resemble the investigative approach). From this perspective, student invention and instructional design and planning are not incompatible. Indeed, we wholeheartedly agree with Gravemeijer's (this issue) main point that "local instructional theories are indispensable for reform mathematics education" (p. 108).

The Realistic Math Education (RME) Model of HLT Development

Gravemeijer (2002, this issue) suggested that the RME model of HLT development embodies a constructivists view and illustrates why student invention and instructional design are not incompatible.

First central tenet.   A central tenet of RME is that the initial portion of an instructional sequence should be experientially real or meaningful to students. Specifically, the key design principle of guided reinvention specifies that students should have the opportunity to rediscover or reinvent aspects of mathematics in a manneranalogous to their original creation. Such meaningful instruction, which promotes autonomy and ownership, is a key characteristic of the investigative approach and is consistent with Dewey's (1963) principle of interaction (Point 3 in Table 3).

However, in another key respect, RME seems more like a conceptual approach than an investigative approach. Gravemeijer's (this issue) discussion seems to suggest that real mathematics does not necessarily mean activities that are purposeful to students. This impression is underscored by the examples of mental arithmetic instruction discussed later. In contrast, a key characteristic of the investigative approach is that a teacher strives, whenever possible, to build on or create a real need for learning and practiciing mathematics (i.e., the emphasis is on real, not merely realistic, mathematical education). Furthermore, RME activities may not take into account key internal factors identified by Dewey, namely the interests and needs of students.

Second central tenet.   Another tenet of RME is that instruction should take into account students' existing (developmental readiness) and "be justifiable in terms of the potential mathematical end points of a learning sequence" (Gravemeijer, 2002, p. 3). The first point is consistent with Dewey's (1963) principle of interaction (Point 3 in Table 3), and the second point is consistent with his principle of ensuring educative experiences (Point 2 in Table 3).

The RME development process.   Gravemeijer's (2002, this issue) description of the development of instructional activity epitomizes Dewey's point that reform efforts or new instructional methods must have their own theoretical, empirical and practical justification, not simply be a reaction to traditional practices (Point 1 in Table 3). Because of this, the RME and other curriculum projects based on a similar development model promise to have a larger and longer-lasting effect than previous reform efforts.

The Example of a Mental Arithmetic Lesson

Like number sense in general, we agree with Gravemeijer's (this issue) general goal of developing "a framework of number relations" that provide a basis "for flexible mental computation," rather than teaching "a set of strategies" (p. 114). Such a view is consistent with Paulos' (1991) advice to "stress a few basic principles and [leave] most of the details to [the student]" (p. 7).

However, the exemplary instructional sequence did not appear to illustrate this perspective well. Based on Beishuizen (1993), the RME team decided to focus on promoting one particular set of strategies (i.e., counting-in-jumps methods) and to dismiss the error-prone strategy of splitting a number into 10s and 1s. But is the latter strategy necessarily more likely to result in errors? Might it not depend on what was taught previously and how this strategy or its prerequisite knowledge was taught? In any case, the RME team's focus on promoting the best strategy seems more consistent with pluralism than with instrumentalism.

Consistent with a pluralist view or the conceptual approach, the team's earlier effort seem to involve imposing an empty number line procedure on students (see Gravemeijer, 2002). Not surprisingly, "problems arose when the teacher posed addition and subtraction tasks by drawing a horizontal empty number line, and by describing transactions in a candy shop, but without acting them out" and that students adapted this representation "primarily as a way of notating, and not as a way of modeling a (mental) activity" (p. 9). The students interpreted the empty number line drawing by the teacher (Gravemeijer, 2002, Figure 6) in two ways. Some students interpreted the diagram for 90 - ? = 88 correctly, as take away 2 (as *90 minus 1 is 89 left, 89 minus 1 is 88 left*), whereas others interpreted it incorrectly as take away 3 (as *take away candy number 90, candy number 89, and candy number 88*). This lack of shared meaning is often the by-product of efforts to impose a method on students.

As mentioned earlier, another common limitation of the conceptual approach is the lack of purposeful activities. Consider three activities described in Gravemeijer (2002). Although the Target Game used in earlier lessons can be somewhat purposeful in the sense that students could view it as a challenge, the Candy Shop activity seems artificial and, from a student's perspective, without much purpose.[1] The Ruler as a Model activity likewise read as a lesson out of a curriculum based on the conceptual approach. The procedures in each part of the instructional sequence appear to be prescribed by the teacher and the activities do not seem to be purposeful from a student's perspective. The instructional sequence described by Gravemeijer (this issue) is an improvement over earlier plans, but the instructional activities still seem more artificial than purposeful.

AN ELABORATION OF HLT
(SIMON & TZUR, THIS ISSUE)

The National Council of Teachers of Mathematics (1989, 2000) has recommended that worthwhile tasks—interesting, challenging, mathematically rich tasks—be a basis for mathematics instruction. Simon and Tzur (this issue) propose an elaboration of HLT that provides a vehicle for selecting tasks and thinking about the learning process. Their efforts build on Dewey's (1963) and Piaget's (1964) views.

Dewey's (1963) principle of interaction and Piaget's (1964) construct of assimilation imply that the selection of tasks must be done carefully and take into account the developmental readiness of students. According to Piaget's *moderate novelty* principle, highly novel information cannot be connected to existing knowledge, cannot be assimilated and, thus, will be uninteresting to a learner. Highly familiar information can be readily assimilated but will not provide accommodations (new learning) or interest. Information that is just beyond a learners' current comprehension can be at least partly assimilated (understood), can provoke accommodation, and will be intriguing to a learner. These principles apply whether an instructor chooses a cognitively complex task or a task geared to promote learning of a specific concept.

Note that Ainley and Pratt's (2002) proposal that teachers resolve the planning paradox by using tasks that are purposeful is an extension of Dewey's (1963) interaction principle. If a teacher takes into account a student's interests and background, ideally they would choose tasks with a purpose or utility.

---

[1]One way to have made the Target activity more purposeful to students would be to transform it into a game. For example, *The Number Goal Game* can be played by two to six children. A large center card (square) is placed in the middle with a number such as 13 printed on it. Each player draws six small squares numbered 1 to 10 from a pile of squares all facing down. The players turn up their squares. On his or her turn, a player can compose two or more squares to a sum equal to the number in the center. If a player had squares 2, 3, 5, 5, 5 and 8, she could combine 5 and 8; and 3, 5, and 5 to make 13. As each solution would be worth 1 point, the player would get 2 points for the round. If the player had chosen to combine 2 + 3 + 8, no other possible combinations of 13 would be left, and the player would have scored only 1 point for the round.

In their review of the HLT construct, Simon & Tzur (this issue) noted that the goals for student learning provide direction for the other components and that task selection and hypotheses about student learning are interdependent. It seems, though, that setting goals is at least somewhat dependent on the issue of what tasks (experiences) are practical to provide and more importantly on what is known about how students' learning progresses with the task and related tasks. Put differently, setting goals (an external factor) is not exempt from Dewey's (1963) principle of interaction. Goals must be chosen with consideration of internal factors, such as what children can be reasonably expected to achieve, as well as other external factors (e.g., societal needs and preparation for more advanced training). Furthermore, goals must be ordered in accordance with developmental readiness and sequence.

Simon and Tzur (this issue) hypothesized that reflection on activity-effect relations is a basis for concept formation and provides a framework for both task selection and understanding the learning process. They noted that this construct is based on Piaget's construct of assimilation and the related process of reflective abstraction.

Consider, for example, the role of manipulatives. Research indicates that simply demonstrating concrete models for students and then requiring them to imitate such manipulative-based procedures does not promote conceptual learning (e.g., Baroody 1989b; Clements & McMillen, 1996; Fuson & Burghardt, 2003; Miura & Okamoto, 2003; Resnick, 1982; Seo & Ginsburg, 2003). This is a key reason why the conceptual approach, although more effective than a traditional skills approach, is not highly effective in fostering mathematical proficiency (Baroody 2003). It is difficult, if not impossible, to impose understanding. Manipulative models are useful if a student reflects on their use and can relate (assimilate) this experience to their existing knowledge. It is an empirical question, but this is more likely to happen if students are asked to use what they know to devise their own solutions than if shown how to use manipulatives. The former method is how manipulatives would be used in the investigative approach.

Simon and Tzur (this issue) did not discuss how the reflection or activity-effect relation might promote a second key way to make connections, namely integration. Whereas assimilation is the process of connecting new information to existing knowledge, integration occurs when a student links two existing but previously isolated aspects of knowledge (Hiebert & Lefevre, 1986).

According to Simon and Tzur (this issue), reflection on activity-effect relations begins with the learners having a goal. They make the much-needed distinction between this and a teacher's goal. The learner's goal is the by-product of a purposeful activity. It provides the motivation to examine the effects of physical or mental actions to gauge whether the goal is being achieved or is achieved.

Some questions the reader may wish to consider about the reflection on activity-effect relations follow.

1.   Is it the (i.e., only) mechanism by which a new concept is constructed?

2.   Although it is possible that the reflective process can be nonconscious, does a concept of number develop with little or no conscious thought, as Simon and Tzur (this issue) argue? Compare, for example, their view with the emerging view in developmental psychology that language plays a key role in concept formation and that labeling collections or seeing them labeled with number words plays a critical role in abstracting number (Baroody, Benson, & Lai, 2003; Mix, Huttenlocher, & Levine, 2002; Spelke, 2003a, 2003b).

3.   Surely, "it is reasonable to claim that learners pay attention to their intentional variations in their goal directed activity" (Simon & Tzur, this issue). However, might not unintended variations that produce surprising results sometimes also promote reflection and learning?

4.   Do all intentional variations produce reflective abstractions?[2]

The example of the equivalent-fractions lesson raised a number of questions readers may wish to consider.

1.   How is the task purposeful to students?
2.   The lesson with its prescribed steps seems consistent with a highly structured discovery approach. Is this consistent with Paulos' (1991) suggestion to focus on conceptual development and leave the procedural details to students to invent?
3.   Is it really consistent with an inquiry-based (investigative) approach recommended by National Council of Teachers of Mathematics?

In brief, the concerns about HLTs raised by Lesh and Yoon (this issue) seem applicable. Is it really necessary to prescribe a single favored path? Is it really necessary to separate this learning from a rich, complex activity?

THE CASE OF COMMENSURATE FRACTIONS
(STEFFE, THIS ISSUE)

Steffe (this issue) offers a detailed look at the development of two children's knowledge of fractions. The tremendous advantage of long-term teaching experiments is that experimenters can get to know participants relatively well and use

---

[2]Consider the case of Felicia (Baroody, 1984). The 5-year-old normally used a verbal counting-all procedure to compute the sums of single-digit combinations (e.g., for 3 + 5, she counted: "1, 2, 3, 4, 5; 6, 7, 8"). When presented with challenge items, Felicia immediately and consistently used either a counting-on strategy (e.g., 5 + 22: "23, 24, 25, 26, 27") or a counting-on-like strategy (e.g., 32 + 6: "31, 32, 33, 34, 35, 36, 37, 38"). However, this intentional variation did not appear to cause refection and conceptual change. When single-digit combinations were reintroduced, Felicia reverted to using counting-all procedures. Moreover, when counting-on was modeled for her with these smaller combinations and she was asked to evaluate the strategy, the girl declared, "You can't do it that way."

this history to better build an explanation of interesting phenomena. This is typically not the case in traditional psychological experiments, in which unexpected behavior often is simply treated as measurement error.

A particularly powerful aspect of Steffe's (this issue) teaching experiment is that he used his history with the participants not only to illustrate his theoretical perspective, but also to make theory-based predictions about their subsequent behavior. This is a welcomed departure from past efforts that focused exclusively on illustrating theory.

Steffe (this issue) wisely notes that, from a constructivist perspective, teachers' and researchers' understanding of students' development must be constantly constructed as they interact with students. Like general scientific paradigms (Kuhn, 1970), these constructions are incomplete and subject to revision or even rejection. Steffe calls this understanding on the part of a teacher or researcher *self-reflexivity*. It is altogether fitting that he cites Simon's (1995) concept of *hypothetical learning trajectories*. Indeed, when based on only a few children and in an area where research and theory are sparse, they can be more accurately a *plausible hypothetical learning trajectory*. In effect, such efforts can serve as an existence proof.

Steffe's (this issue) concept of self-reflexivity is analogous to what Sagan (1997) described as the characteristic of any good scientist: ruthless self-assessment or self-criticism. A key component of this process, according to Sagan, is a constant effort to propose and consider plausible alternative hypotheses or explanations. This is especially important with case studies in which the goal is a detailed understanding and explanation and experimental controls of threats to internal validity such as history and testing effects are not used or possible.

Based on the evidence presented, it does appear that overall Jason has a more complete and flexible understanding of fractions.[3] However, there are other factors that can help account for some of the differences between Jason and Laura and raise questions about Steffe's (this issue) conclusions.

1. Laura Seemed Less Adept at Understanding the Requirements of New Tasks (i.e., Procedural Competence Could Have Been a Confounding Factor)

Consider Laura's difficulty in Protocol II. Steffe (this issue) concluded that her uncannily accurate estimates of 1/10 in Protocol I was the product of a localized and

---

[3]Curiously, Laura was uncannily accurate in partitioning the sticks into a specified fraction. However, she was considered to have a conceptually less mature conceptual schema because, in part, she did not check her estimates. An alternative conjecture is that Laura's understanding of equal partitioning and an effective halving strategy for applying this concept enabled her to partition sticks in a highly accurate manner. For example, to represent 3/4, she may have mentally partitioned the stick into halves and then each half into halves. Furthermore, unlike Jason, Laura may have been confident in her knowledge and skill and so felt no need to check her partitioning efforts. Unfortunately, the interviewer did not ask her how she determined her answers so accurately or why she felt no need to check her answers (i.e., did not explore the processes underlying her performance).

relatively nonadvanced conceptual understanding of fractions. This incomplete understanding was inadequate for other fractions and the more advanced tasks used in Protocol II.

Unfortunately, Steffe (this issue) appears to have based this conclusion on follow-up testing that involved changing both the fraction involved and the nature of the task. Specifically, the new task in Protocol II involved both eighths and a new constraint, namely, using only one mark to perform the equal partitioning. An alternative hypothesis for Laura's weaker performance in this situation is that she did not understand the constraint of the task and, thus, could not use her conceptual understanding of equal partitioning to devise an appropriate procedure. (Indeed, these readers had to read the description of the new task carefully several times to understand what the new task required.) We did not find it surprising that Laura, as the protocol makes clear, was confused by the new task.[4] Thus, the following questions remain unanswered: How would Laura have responded to the request to find 1/8 with the familiar task described in Protocol I? How would she have responded to similar requests if task demands had been presented more clearly or she had otherwise been helped to understand the task demands (constraints) of the new task? What if the researcher/teacher had posed the problem or task differently? In brief, changing both the fraction and the nature of the task simultaneously confounds the interpretation of Laura's discrepancy, and inferring conceptual knowledge from the existence or absence of an advanced procedure can be dangerous (cf. Baroody, 1985).

2. Order Effects and Affective Factors Are also Possible Confounding Factors

A common criticism of long-term case studies or teaching experiments is that it is difficult to disentangle order effects. What if Protocol II had preceded Protocol I (i.e., what if the order of the tasks had been switched), how would Laura have responded to the former?

---

[4]Steffe (this issue) considers "the construction of learning trajectories in the context of 'the idea of worlds being constructed, or even autonomously invented, by inquiries who are simultaneously participants in the same world' (Steier, 1995, p. 71)." However, autonomously can imply no interaction or no effect on each other and this is surely not the case. The confusion by Laura about the requirements of the Protocol II task and her inability to understand Jason or the interviewer in other situations underscores the need for shared meaning.

Furthermore, like earlier units-type research, different conceptual levels are inferred from success on more advanced tasks. For example, the construct Steffe (this issue) calls the *splitting operation* entails the composition or union of partitioning and iteration schemes. He conjectures that a child with such an operation is capable of featuring a hypothetical stick that for 1/5, for example, can be repeated five times so that it is equal in length to a real stick. The operational definition for this construct is Protocol II, where a child is allowed to draw only one line on the original stick and then use this segment to create a new stick (5 times long in the case of 1/5) by making copies of the segment. However, a child in Protocol I might mentally imagine the real stick partitioned once to make a fifth and then use this mental image of a fifth to construct an imaginary stick next to the original.

Interestingly, Jason encouraged Laura to tackle the task first in Protocol II and in a number of other situations. After Laura responded incorrectly and after hearing the teacher's feedback, Jason then responded correctly. This trend suggests that Jason in this and perhaps other situations also did not initially understand the demands or constraints of the teacher's task or question, used Laura to obtain clarification, and as a result was able to respond correctly. If so, Jason's strategy is clever, and clearly he seemed to profit more readily from the feedback on Laura's mistakes. The latter could well have been due to a more complete conception of fractions. However, affective factors could also have played a role. For example, while Laura is dealing with the negative affect of again being wrong, Jason can focus on figuring out the teacher's intention.

3. Another Source of Confusion and, thus, Another Possible Confound Was Differences in How to Define the Whole in the Fraction of a Fraction Question in Protocol III

After Laura's partner had created a model of 3/4, she was asked to use the computer tools to find another way to make fourths. Laura's solution was "you can make it smaller." She proceeded to copy one of the 1/4 segments in her partner's model, divide it into fourths, and color in three of the equal segments.

Laura's solution appears to have been to create a scaled down and similar model of her partner's model. If so, she viewed the copied 1/4 segment differently than her teacher. To her, she created a new whole that was 1/4 the size of her partner's whole and viewed its three shaded portions as a fraction of this new whole, a whole now independent of her partner's whole. Steffe (this issue), though, still viewed the copied 1/4 as still a part (1/4) of her partner's whole and, thus, its three shaded parts as 3/4 of 1/4. Both interpretations, of course, are valid. (This is one reason why students and teachers alike should explicitly define the whole when working on fraction problems; Baroody & Lai, 2002.)

A possible difficulty is that Steffe's (this issue) questions (about the fractional name of the three shaded parts in comparison to the partner's larger whole) required Laura to flexibly switch what she defined as the whole. This unexpected switch in the requirements of the task could have caught her off guard and help account for her erroneous response.[5] In such a situation, she could have drawn on her existing knowledge to manufacture an answer. Perhaps she drew on experience with base-10 blocks to quickly generate the estimate of 1/10 (without documentation of her classroom experiences, it is not possible to evaluate this conjecture). In

---

[5]It is our experience that even preservice teachers frequently do not clearly and explicitly define wholes in fraction problems and have considerable difficulty switching among them (e.g., recognizing that the 2/3 in 2/3 × 4/5 is a fraction of the 4/5 and that the 4/5 and product 8/15 are both fractions of the same wholes; Baroody & Lai, 2002).

short, Laura's difficulty might not have reflected a weak conceptual understanding of fractions but a difficulty in applying her knowledge when a question based on a different premise (different definition of the whole) than she held was posed.

4. The Tasks Involve Flexibly Switching Between Two Conceptions of Fractions (Two Meanings of Rational Numbers)

The fair-sharing analogy (e.g., sharing a stick fairly among 15 people) used by the teacher entailed a quotient meaning of fractions. The fractional models (set models, the region model, fraction strips, or the number line model) Steffe (this issue) seems to oppose are commonly used to model a part-whole meaning of fractions.

We agree with Steffe's (this issue) implied suggestion that instruction should begin with problems involving a quotient meaning. However, we disagree that equal-partitioning experience (the big idea underlying both quotient and part-whole meanings of fractions) should begin with fair sharing (quotient) problems that involve a continuous quantity, such as the length of a stick. For several reasons, it makes sense to begin such experiences in kindergarten by fairly sharing collections or discrete quantities and to then introduce fair-sharing problems that involve continuous quantities (see Baroody with Coslick, 1998, for a detailed explanation).

Unlike Steffe (this issue), we believe that it is relatively unimportant what particular model is used. What is important is that children understand (a) how the big idea of equal partitioning applies to the quotient and part-whole interpretations of fractions and (b) the intimate connection between these two meanings of fractions. Consider Steffe's length model. After a child is asked to share a length, for instance, among 15 children (a quotient meaning), determining the solution of the problem (the size of each person's share) requires reinterpreting the stick in terms of a part-whole meaning (Baroody with Coslick, 1998). In other words, the child must view the size of one person's share in relation to the whole partitioned into equal size pieces, namely as one of 15 equal size parts. Note that other manipulatives, such as circular or rectangular region (area) models, could be used to model quotient problems and then represent the solution as a part-whole fraction (e.g., How much of a pizza would each of three children get if they shared two pizzas fairly among them?).

Although Jason appeared to switch easily from a quotient to a part-whole meaning, Laura apparently did not. The reason for this is not entirely clear. Perhaps her thinking was tied more closely to the formal part-whole instruction received in class than was Jason's (as implied by Steffe, this issue). Perhaps her understanding of a part-whole meaning was not well established. Perhaps she did not understand the interviewer's questions, which did not explicitly make clear which of the two different meanings of fractions was of interest. Like most people, perhaps she had never explicitly considered how the two meanings are related. The reader is left to wonder how Laura would have responded if prompted to consider explicitly the quotient and part-whole meanings and how they are related.

COMPOSITION OF GEOMETRIC FIGURES
(CLEMENTS, WILSON, & SARAMA, THIS ISSUE)

Clements, Wilson, and Sarama's (this issue) article provides an invaluable service by illustrating how the big idea of composing and decomposing applies to the topic of geometric shapes, specifically, plane figures. It also illustrates well the value of case studies (e.g., Sarama, Clements, & Vukelic, 1996) in uncovering interesting developmental possibilities or serving to reveal what is possible. The Developmental Progression section illustrates the value of thinking about development along a dimension (cf. Lesh & Yoon, this issue). The progression from relatively little and global knowledge to relatively complete and detailed knowledge is useful for researchers and educators alike—as long as it is not reified.

The article further illustrates how a HLT approach can be done carefully and thoroughly to develop both theory and practical instructional and assessment materials. Clements et al.'s (this issue) involvement of scholars from a variety of fields and practitioners is a model for collaborative efforts.

One main concern is the reliability and validity of Clements et al.'s (this issue) assessment instrument. For example, both Task 1 and Task 2 assessed the Shape Composer level, and both Task 1 and Task 3 did so for the Substitution Composer level. What was the reliability in each case? That is, how well did the participants' performance on Task 1 match with their performance on Task 2 or Task 3? For example, theoretically, a child scored as a Shape Composer on Task 1 would also be scored as such on Task 2.

Furthermore, the authors could have checked their developmental progression statistically by using a classic statistic, such as a Guttman scale or trend analysis, or newer order-sensitive statistical tools, such as a latent class analysis.

A second more minor concern is that not all levels appeared to be tied to a conceptual advance. Cements et al. (this issue) did make an effort to relate their levels to van Hiele (1986) levels. For example, Levels 0 and 1 (Precomposer and Piece Assemble) were tied to the first van Hiele level. However, what is the difference between Levels 0 and 1 other than procedural knowledge? What enables Level 2 children (Picture Makers) to form pictures in which several shapes play a single role? The remaining levels, though, are tied to or extend the van Hiele levels.

COGNITIVE-BASED ASSESSMENT
(BATTISTA, THIS ISSUE)

We emphatically agree with Battista's (this issue) premise that the development of effective assessment tools depends on linking it to developmental research. Conversely, cognition-based assessment (CBA) enriches HLT by elaborating on how to assess and define students' concept levels and on how to help students move from one concept level to a more sophisticated one. By using CBA, a teacher can gauge a student's current level and decide on an activity to promote his or her development. Battista correctly points out several key advantages of learning trajectories. One is that they specify both what students can and cannot do and the cognitive obstacles to learning progress.

Another advantage is that, unlike most previous efforts to define learning sequences, HLT specify the mental processes for a level and for progressing to a higher level in sufficient detail that the jumps in sophistication are small enough to be made meaningful and with relative ease. What Battista (this issue) describes as students' "zone of construction," then, is analogous to Vygotsky's (1978) "zone of proximal development" and to Piaget's "moderate novelty principle." Thus, it provides educators with crucial information about developmental readiness.

Battista (this issue) noted that the first of three critical components of CBA is a description of core ideas and processes. His view of core ideas is related to, but not identical with, the concept of "big ideas." The core idea for measuring area and volume is "understanding how to meaningfully enumerate arrays" of square units and cube units, respectively. Battista then describes various core processes that enable children to construct this core idea. The discussion is highly complicated and as such is probably not practical. Moreover, it seems to confound procedural knowledge with conceptual knowledge.

Consistent with Paulos' (1991) advice noted earlier, we recommend that teachers focus on big ideas and leave the procedural details to students. The big idea underlying area, volume, and all other measurement concepts is equal partitioning. A continuous quantity by definition does not have distinct components and so cannot be quantified by counting, as can be done with a discrete quantity. However, if a continuous quantity can be partitioned into equal size parts, then the number of these equal size parts (units) can be counted. Once children grasp this big idea, they should be able to figure out for themselves (or with minimal guidance) how to measure any continuous quantity.

In our view, then, the big idea of equal partitioning should provide a key basis for measurement HLTs. Consider, for instance, Bill's work used to illustrate Level 2 (e.g., Battista, this issue, Figure 5). In our view, a teacher should focus on helping this student coordinate the big idea of equal partitioning with the concept of a measurement unit (i.e., on building conceptual links and knowledge). Once Bill understands that measurement is simply the process of partitioning a continuous quantity such as area into equal parts (units) so that the parts can be counted, he can be left to invent a procedure for counting the parts. (Equal partitioning, in effect, transforms a continuous quantity, which cannot be quantified by counting, into a discrete quantity, which can be quantified by counting.) This approach simplifies the knowledge demands on teachers and is, thus, more likely to be successful. Although the big idea of equal partitioning (a continuous quantity into countable units) could theoretically allow children to devise various measurement concepts in any order, it could facilitate meaningful learning (reinvention) of measurement formulas for area and volume if students first learned how to measure lengths, then area, and finally volume. For instance, once students understood the concept of area and reinvented the formula of $l \times w$ as a shortcut for more concrete procedures for gauging it, volume problems could then be introduced. By applying the big idea of equal partitioning and reflecting on their concrete models, students can recognize that volume is simply so many layers of a given area. For instance, a 4 cm long by 3 cm wide by 2 cm deep is two layers of 4 cm × 3 cm or 12 cm² × 2 cm or 24 cm³. Summarizing this discovery using the shorthand of algebra leads to $V = A \times d$ or $l \times w \times d$.

The HLT suggested by the previous discussion could be: (a) understands equal partitioning with discrete quantities; (b) can apply equal partitioning to linear quantities; (c) can use the previous goal to measure length; (d) can apply equal partitioning to gauge the area of plane figures; (e) for rectangles, recognizes the shortcut for counting the number of squares (square units), namely, multiply the length and the width of a rectangle; (f) summarizes the previous discovery algebraically; (g) can apply equal partitioning to solid figures; and so forth. Note that understanding the big idea of equal partitioning can theoretically lead to devising the formula for volume and other plane and solid figure in the same way described for area (Steps d, e, and f).

Likewise, Katy's work, illustrated in Battista's (this issue) Figure 3, indicates she does not understand that the area of a rectangle must be decomposed into equal size parts. Helping her understand the big idea underlying area in particular and measurement in general is a clear and straightforward goal. Understanding this big idea should also reduce her and other students' tendency to double count squares.

In brief, focusing on big ideas makes a teacher's already complicated life significantly more manageable. For example, it can simplify the second critical component in CBA, namely knowing for each core idea, the research-based descriptions of the cognitive constructions students must make to understand an idea. Parenthetically, it would be more accurate to describe the third critical component as follows: "For each core idea, coherent sets of assessment items that enable educators to investigate students' cognitions and *to estimate as* precisely *as possible* ~~locate~~ students' positions in the 'constructive itineraries' typically taken in acquiring competence with the idea" (changes indicated by italics and strike through). In a similar vein, Battista's conclusion about assessment would be more accurate if stated: "CBA can carefully **estimate** ~~examine~~ *the ex-act nature of students' cognitions* **with relatively good precision**" (changes indicated by bold and strike through).

Finally, we agree that educators need to know both the *if* and *what*. This distinction is similar to Ginsburg's (1977) distinction of assessing both product (whether a child can produce a correct answer) and process (how a child arrives at an answer).

EVOLVING COMMUNITIES OF MIND
(LESH & YOON, THIS ISSUE)

Lesh and Yoon (this issue) ask a fundamentally important question: Do ideas really develop along a trajectory? After noting several difficulties with linear or ladder-like sequences and even the more complex multiple-paths or branching-tree models, they suggest that in many cases the answer is no. We agree that a model of development should entail both sequential learning, including linear or ladder-like paths in some cases, and interconnecting or even web-like features. Unlike the conceptual-like approach some proponents of learning trajectories (e.g., Gravemeijer, this issue; Simon & Tzur, this issue) seem to advocate, Lesh and Yoon's models and modeling perspective is consistent with the investigative approach in many ways. In the following discussion we evaluate, in turn, Lesh and Yoon's comments on trajectories and their models and modeling perspectives.

Learning Trajectories

Lesh and Yoon (this issue) correctly note that models of learning trajectories focus on domain-specific development, not the development of general cognitive structures of interest to Piaget and his colleagues. This shift in focus is part of a larger post-Piagetian trend over the last three decades in development and educational psychology—a shift that has made developmental research far more useful to educators.

Linear or ladder-like sequences. Lesh and Yoon (this issue) seem to imply that the linear development along a single dimension or continuum, such as concrete-abstract or specific-general suggested by ladder-like sequences, is an over simplification of a complicated process. In effect, they reject the dualistic-like view (see the first entry in Table 1) that development strictly follows a sequence of increasingly sophisticated stages of understanding. For the most part, they are probably correct; such models are almost surely an over simplification of development.

Even so, ladder-like sequences can be useful models. For instance, they provide a starting point for empirical testing and theory elaboration that can lead to the branching tree or even more complicated models of development. They can also serve as a good starting point in educating pre- and in-service teachers. After all, the basic idea behind ladder-like sequences—that children gradually move to a more and more complete and accurate understanding of concepts (i.e., that concept learning is not an all-or-nothing process)—is a fundamentally important one for an educator.

Lesh and Yoon (this issue) further criticize ladder-like models because they imply neat topic boundaries and that students must be proficient with one idea before moving on to another. This is, of course, not always the case. There is no developmental reason that addition should be introduced before subtraction in first grade and that multiplication should be postponed until third grade. Children informally understand both addition and subtraction before they enter kindergarten and the formal instruction of simple subtraction, at least, could be done simultaneously with addition. Given children's difficulties with subtraction of numbers larger than 5 and the intimate relation between (repeated) addition and multiplication, it makes sense to introduce the latter instead of the former in the last half of first grade or in second grade (Wynroth, 1986). In brief, we agree that strict adherence to the scope-and-sequence charts found in typical textbooks is frequently not productive.

Nevertheless, instruction should not be haphazard either. It makes sense to introduce decimals after children have some understanding of place value, equal partitioning, fractions in general, and decimal fractions in particular. Indeed, with a basic understanding of these related concepts, children are in a good position to reinvent decimal notation themselves (Baroody with Coslick, 1998).

In summary, it is important for educators to realize that developmental readiness is an important factor in learning and that care should be exercised in considering what topics, activities, and problems are chosen and in what order. Again, ladder-like sequences, particularly those that help spell out how different sequences are related is an important starting point for educators. It provides a practical way of initially planning and organizing instruction, whether its actual implementation proceeds along a predicted path or not.

Multiple-path or branching-tree trajectories. Lesh and Yoon (this issue) accurately point out that branching-tree models are often a more accurate description of development, because such models explicitly recognize that social and individual factors create multiple paths to understanding. However, they seem concerned by the pluralistic premise of these models (see the second entry in Table 1), namely, that there are multiple possibilities but ultimately only one best or "politically correct" path. Although this criticism is valid in some cases (see Simon and Tzur's example of equivalent fraction in this issue and Gravemeijer's, this issue, example of mental addition), it does not apply to branching-tree models. For example, Battista's (this issue) Figure 1 does not suggest a favored path and is consistent with the instrumentalism view (see the third entry in Table 1), namely, that there are many possible and equally effective paths. Also consistent with this philosophical view, Clements and Sarama discuss "a 'best-case' instructional sequence." Note their careful use of the indefinite article *a*, which implied the instrumental view, rather the definite article *the*, which would have implied the pluralistic view.

Furthermore, although some models of development could be more consistent with a particular philosophical orientation than others, a pluralistic belief, for instance, is held by individuals and not inherent in a model. For example, Lesh and Yoon (this issue) observed: "Negative effects of teacher-imposed (or community-imposed) political correctness are being seen consistently in Purdue's *Gender Equity in Engineering Project* (Ashmann, Zawojewski, & Bowman, 2003), which focuses on students' performance in model-eliciting activities similar to the one that will be described in this article." Put differently, it is not fair to criticize branching-tree models for the (pluralistic) belief of some of its proponents and users when some of the proponents and users of model-eliciting activities hold the same belief.

Finally, there are numerous cases in which observations or research indicate that an instructional sequence promotes development or learning more effectively than some others. Our own case study work indicates that some parents begin number instruction with counting objects. This appears to be confusing to children because number words are used in two different ways. During the counting process, they are used as ordinal terms (to specify the order of the items counted). At the end of the counting process, the last number word used also has a cardinal meaning, namely, it specifies the cardinal value of the collection (the total number of items in the collection). Understanding this cardinality principle is fundamental to functional or meaningful object counting. Children's confusion about this switch in number meaning (the absence of a cardinality principle) can be manifested by, for example, their recounting the collection when asked "How many?" Our observations indicate parents' efforts to impose counting on children often leads to learning the object counting process by rote and its nonfunctional, mechanical, or error-prone application.

A more optimal instructional route suggested by research (e.g., Baroody et al., 2003; Starkey & Cooper, 1995; von Glasersfeld, 1982) is to first encourage verbal number recognition of small collections. For example, by initially labeling various examples of single instances "1" and various examples of pairs "2" and larger collections (nonexamples) as "not 1" or "not 2," children can abstract a concept of oneness and twoness. Once this is achieved, a child can in a similar way be helped to abstract threeness and then fourness.

With this basis, counting instruction should make more sense to young children. If a parent models counting a collection of three items, for example, it is far more likely a child will understand the switch in how the last number word is used. Specifically, when the parent models "1, 2, 3, see 3," or "1, 2, t-h-r-e-e," the child who has already used verbal number recognition to see that there are three things in the collection can better understand why his or her parent repeated or emphasized the word "three" and, thus, is more likely to abstract the cardinality principle. Is this the best way to help children understand? Perhaps, perhaps not, but it seems to be a more effective path than what many parents and preschool teachers currently use.

Trajectories in general.   According to Lesh and Yoon (this issue), many of the problems with trajectories stem from the assumption that a stage of understanding is stable across applicable tasks. They argue that this assumption is contradicted by the enormous literature that shows task variables influence task difficulty and overlooks theoretical constructs, such as Piaget's (Piaget & Beth, 1966) *decalage* for explaining inconsistent performances across different but conceptually related tasks.

There are at least two problems with Lesh and Yoon's (this issue) criticisms of learning trajectories. One is that the mountain of evidence they mentioned includes a great deal of research that focused on knowledge learned by rote. Such knowledge typically does not transfer to even moderately different tasks. Furthermore, their criticism is something of a straw man. Not all or even most proponents of learning trajectories believe that conceptually based knowledge will automatically transfer. That is, the more informed proponents of learning trajectories recognize the possibility of task effects.

Indeed, in the end, Lesh and Yoon (this issue) do not themselves seem to dismiss branching-tree models or even ladder-like trajectories. Their basic argument seems to be that these models do not provide a complete picture of development. With this, we do agree.

A Models and Modeling Perspective

We first evaluate the evidence adduced by Lesh and Yoon (this issue) and then attempt to put their model in perspective.

The evidence.   "To challenge the assumption that ideas develop along a trajectory," Lesh and Yoon (this issue) examine a single example that supposedly does not develop in such a fashion. What the authors intend by this argument is not clear. They could have been proposing that their existence proof of a case of web-like learning disproves that all ideas develop along a trajectory. Although there could be those who need to be convinced of this, we suspect most readers would not contend otherwise. Perhaps the authors were challenging the assumption that only some ideas (rather than all) develop along a trajectory. If so, then a single counterexample is a start but not sufficient to support their claims. Perhaps what the authors meant to say was: To demonstrate that a body of ideas can develop in a nonsequential manner.

To their credit, Lesh and Yoon (this issue) chose as participants inner-city remedial students. That is, they stacked the deck against their finding success with their teaching approach. Lesh and Yoon imply that the results reported were representative of the class. This stands in contrast with research that indicates children with special needs, overall, do not seem to benefit from standards- or inquiry-based instruction (e.g., Baxter, Woodward & Olson, 2001; Boaler, 1998; Fuson, Carroll, & Drueck, 2000; Sowder, Philipp, Armstrong, & Schappelle, 1998; Woodward & Baxter, 1997). The work by Bottge and colleagues (e.g., Bottge, Heinrichs, Chan, & Serlin, 2001; Bottge, Heinrichs, Mehta, & Hung, 2002) is only somewhat more optimistic. They found that a project-based approach helped low achieving middle school students significantly improve their problem-solving performance but had little impact on content achievement. In light of these results, supporting data about the success of other teams would have made Lesh and Yoon's (this issue) case more convincing. Readers could also wonder: What special supports could have helped those less successful students?

Lesh and Yoon (this issue) repeatedly state that their approach can produce major shifts in students' thinking in 60 to 90 min. There are two problems with their evidence.

1.   The Quilt Problem entailed a preliminary session to discuss a relevant newspaper article. The problem-solving session took 90 min by itself, and the follow-up discussion took another 90 min. The latter presumably helped some students in the class make a shift in thinking (or otherwise was not an effective use of class time). This does not even count the time spent on similar scaling problems done earlier. There are, of course, many worthwhile tasks that do not require more than 90 min, but Lesh and Yoon seem to underestimate the complexity and time required by many or even most worthwhile tasks.

2.   The significant change in thinking involved one student's insight that the Quilt Problem involved, like a previous problem, scaling up. This insight or transfer is not unimportant. However, this example involves a qualitative change in thinking about a specific task—a local application of a broader insight about scaling up that came from working on the previous problem or perhaps earlier. As such, the example is somewhat disappointing.

The report by Lesh and Yoon (this issue) provides tantalizing clues of the broader implications of their research data, but ultimately, readers may wonder about its significance. Did students retain their conceptual advances and transfer their learning to other different problems or contexts? Were there individual differences in performance and why? How could these differences be addressed? How would the students have responded without prior scaling up activities?

Perspective.   We strongly agree with Lesh and Yoon's (this issue) recommendation that a basis of instruction should be model-eliciting activities. The use of open-ended tasks that are purposeful, meaningful, and inquiry-based is important for making mathematics engaging, thought provoking, and effective. This approach epitomizes the investigative approach in many ways and seems consistent with Paulos' (1991) advice to focus on few basic principles and to leave most of the details to students.

Our endorsement of models-and-modeling perspective comes with four qualifications, qualifications that Lesh and Yoon (this issue) themselves might endorse or not.

1.   Although the model-eliciting activities recommended by Lesh and Yoon are clearly useful and valuable, they are only one type of worthwhile task for mak-

ing instruction purposeful, meaningful, and inquiry-based. Put differently, a variety of tasks can be used to implement the investigative approach effectively.

2.   The design feature of documenting solutions for a "realistic 'client'" is clearly an improvement over writing justifications for solutions as an after thought. Although this does create an artificial purpose for the writing task, ideally teachers would use tasks that have a real purpose to students.

3.   Lesh and Yoon (this issue) do not indicate how they gauged whether the students were developmentally ready for the model-eliciting activity. Whether intentionally, they create an impression that developmental readiness was left to chance. If so, then this is a way in which the models and modeling perspective resembles the laissez-faire problem-solving approach described in Table 2.

4.   Lesh and Yoon (this issue) do not discuss the role of teachers during model-eliciting activities or in follow-up class discussions. Indeed, they explicitly state, "we explain how these activities reliably elicit observable idea development without direct teacher guidance." The result is an impression that teachers play little or no role in such an approach other than to provide students with model-eliciting activities and the opportunity to explore them on their own and discuss them with the class. A reader could well wonder: What would have happened if Ann, like her two group mates, did not know how to use a ruler to make measurements? If our impression is accurate, then this is another way models and modeling perspective resembles the laissez-faire problem-solving approach.

Consistent with the investigative approach, we believe that teachers need to play an active, if indirect, role in guiding learning and cannot simply present a series of activities haphazardly (Baroody with Coslick, 1998). Although teachers should give students the leeway to devise their own strategies and construct their own understandings, they can play an important role in encouraging their development by promoting (social and cognitive) conflict or doubt and, thus, reflection, discussion, and insight.

Consider, for example, Ann's repeated efforts to measure separate pieces of the quilt and add their lengths, each time obtaining a measure inconsistent with the described width of the quilt. Here a teacher could have posed to the group or the class the question of why this discrepancy kept occurring. With any luck, the discussion could have led to the insight that any measurement is essentially an estimate and that the discrepancy was due to the cumulative effect of measurement errors. This insight would have helped the students understand something fundamentally important about measurement and, perhaps, saved Ann and her group precious time and effort.

Teachers should also use what they know about their students to plan a series of developmentally appropriate, worthwhile tasks. Dewey (1963) learned from his progressive school experience (essentially a laissez-faire problem-solving approach) that experiences (activities) per se were not necessarily educational. He concluded that truly educational experiences (activities) occurred if external factors (e.g., the nature of an activity, the classroom, peer and teacher input) meshed with internal factors (e.g., interest, needs, and developmental readiness). Ensuring that students are developmentally ready for any given model-eliciting activity requires considering learning trajectories at least to some extent. Of course, even the best-laid plans should be considered tentative and subject to revision. As Eisenhower noted, before a battle, plans are everything; once the battle begins, plans are nothing. Hyperbole aside, our point is that teachers should flexibly change course, for instance, to take advantage of questions and other "learning moments."

CONCLUSIONS: THE ROLES OF BIG IDEAS

We conclude with how big ideas can serve to make HLTs more practical and how they can extend the theoretical model or analogy proposed by Lesh and Yoon (this issue).

A Practical Consideration

In general, we agree with Lesh and Yoon's (this issue) concern about how HLTs are sometimes perceived and used. The HLTs proposed by Steffe (this issue) and by Battista (this issue) are so highly technical and complicated it is unlikely that practitioners would find them helpful. First, a teacher would have to invest a significant amount of time to understand these HLTs. Second, gauging the status of each child's development, implementing and monitoring the instruction of each (even if grouped), and then assessing the progress of each student would require an enormous amount of work on the part of teachers. The HLTs proposed by Simon and Tzur (this issue) and Gravemeijer (this issue) would be more comprehensible to practitioners. However, although many might welcome the step-by-step procedure for achieving the ultimate goal in each case, both of these HLTs seem overly prescriptive and inconsistent with an inquiry-based investigative approach. They also seem inconsistent with Paulos' (1991) advice to focus on key concepts and leave the details to students.

HLTs could be made more comprehensible and useful to practitioners if they focused on how big ideas evolve. A key reason for this is that big ideas provide a relatively strong and general basis for adaptive expertise—meaningful learning that can be applied to moderately novel tasks as well as familiar tasks (Hatano, 1988, 2003). Understanding the big idea of equal partitioning can empower students to understand a wide range of concepts and to invent (or reinvent) a variety of procedures (see Table 4). This is why Paulos' (1991) previously mentioned advice makes practical sense.

TABLE 4
Focusing on Big Ideas

Teachers should focus on helping students discover and understand big ideas—key ideas that underlie numerous concepts and procedures across topics. If students understand the big ideas, most will be able to rediscover or reinvent the principles, properties, and procedures central to elementary arithmetic and geometry, including the renaming procedures, the commutative and distributive principles, and area formulas. Understanding big ideas can help students understand the rationale for specific methods (e.g., procedures and formulas), adapt them to meet the challenge of new problems or tasks ("adaptive expertise"), and see how various concepts and procedures are related. This can help students see that mathematics is a system of knowledge. This in turn can make learning diverse ideas and procedures much easier.

Consider, for example, the big idea of equal composition or decomposition (partitioning): Equal size parts can be used to compose a whole and a whole can be divided into equal parts. Equal partitioning can be related to children's informal experience of fair sharing. Fair sharing (equal partitioning) can provide a conceptual basis for such diverse concepts as:

- Unit principle—any number can be expressed as the sum of units (e.g., $5 = 1 + 1 + 1 + 1 + 1$)
- even number—an even number of items can be shared fairly by exactly two people
- division—either partitive (a whole shared fairly among a certain number of people) or quotitive division (a whole divided into equal size shares)
- fractions—both a quotient meaning (e.g., 3/4 can be viewed as: Sharing three candy bars among four people, what is the size of each person's share?) and a part-whole meaning (e.g., 3/4 can be viewed as: What part of the whole candy bar is three of four equal size pieces?)
- measurement—a continuous quantity such as length or area can be subdivided into equal size parts or shares (units), which can then be counted
- mean—to find what a typical share size (score) would be if everyone had the same size share, combine all shares (add the scores), and then divvy up the total fairly among the number of people (divide by the number of scores)

Theoretical Extensions

The genetic inheritance tree model proposed by Lesh and Yoon (this issue, see their Figure 8) makes a great deal of sense but can be improved with a few modifications such as incorporating the concept of big ideas. Their model combines the useful aspects of both linear or ladder-like trajectories and multiple-path or branching-tree trajectories. It also explains why some instances of development do not involve sequential learning. The genetic inheritance tree model squares well with the conventional wisdom that the depth of understanding depends on the extent to which knowledge is interconnected (e.g., Ginsburg, 1977; Hiebert & Lefevre, 1986).

What the genetic inheritance tree model does not do, though, is describe explicitly how big ideas can play a key role in meaningful learning. To take into account this and other aspects of learning, we offer an extension of Lesh and Yoon's (this issue) model, which we call *big ideas model of trajectory learning* (see Figure 1).

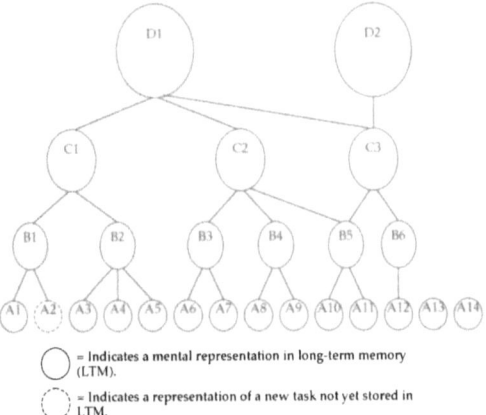

FIGURE 1   An extension of the genetic inheritance tree model (e.g., Lesh & Yoon, this issue) models—the big idea's model.

The processes of integration and assimilation.   We feature big ideas as a key mechanism for linking or integrating different branches of the inheritance tree. For example, a really big (overarching) idea (D1 in Figure 1) can serve to connect several big ideas in different content areas (e.g., C1, C2, and C3 in Figure 1). A big idea can do the same for topics within a content area (e.g., B1 and B2 in Figure 1). A modestly big idea (e.g., B1 in Figure 1) can connect different problems or tasks within a content area (e.g., A1 and A2 in Figure 1). A concrete example of this was Carla's insight (described in Lesh & Yoon, this issue) that scaling can be applied to the Quilt Problem as well as the Bigfoot Problem.

Unlike Lesh and Yoon (this issue), then, we believe that the scale of an idea affects its stability and influence across contexts. We agree that their distinction between "a stages *N* child" versus "the levels of understanding on a specific task" (e.g., also see Battista, this issue; Clements, Wilson, & Sarama, this issue) is an important one. However, we believe there is a place in developmental/instructional theory for ideas that are stable across structurally similar tasks or contexts. Big ideas of various degrees serve this purpose. Really big ideas provide the greatest opportunity for adaptive expertise or transfer; big ideas, less so; and so forth.

Consistent with the Lesh and Yoon's (this issue) metaphor of ideas functioning like a biological organism, the nodes in Figure 1 can be viewed as organic and capable of generating new connections. Carla's previous experience with the Bigfoot Problem (Node A1) and her more general understanding of scaling (Node B1) enabled her to see a commonality between the previously solved problem and the new Quilt problem (Node A2). This allowed her to assimilate the new problem in terms of her existing knowledge and apply and adapt her existing knowledge of scaling to solving the new problem.

The process of integration would entail recognizing a commonality between two existing but previously unconnected nodes and growing links to them. The bigger, the more interconnected, and more active the idea, the more likely a commonality will be found between previously unconnected lower level nodes.

Big ideas and weak versus strong schemas.   What causes one node to be relatively active and thus more likely to grow new connections (via assimilation or integration) than a less active node? The activity of a node can be likened to the strength of a schema (see Table 5). A weak schema represents local knowledge or routine expertise that can be applied in a narrow, nonlogical fashion. A strong schema represents meaningful (principled) knowledge or adaptive expertise that can be applied broadly and logically. Thus, a really big idea that is well understood (represented by a strong schema) is able to survey existing lower level nodes or a new activity and more likely to make or grow a connection. Thus, such an idea affords the application of more cognitive resources (adaptive expertise) when solving a new problem or learning new material than do lower level nodes, particularly those represented by a weak schema. This is why focusing on big ideas can enable students to sort out the details for themselves.

The relation between a node representing a big idea and lower level nodes is not a one-way process. By reflecting on two lower level ideas, it is possible for a student to recognize a commonality and construct (i.e., abstract) a big idea. This big idea could, in turn, find additional applications, as discussed in the previous para-

TABLE 5
Continuum from Weak to Strong Schema

Weak Schema	Relatively Weak Schema	Relatively Strong Schema	Strong Schema
Entail generalizations local in scope			Entail generalizations broad in scope
Low standards of internal (logical) consistency			High standards of internal (logical) consistency
Comprehension is precedent-driven			Comprehension is principle-driven
No logical basis for a priori reasoning; predictions are looked up			Principled (logical) basis for a priori reasoning; predictions are derived
Informal thinking is intuitive, infra-logical, and implicit	Informal thinking is principled but still unsystematic	Informal thinking is principled and systematic	Formal thinking uses formally defined and explicit principles

graph. Unlike Lesh and Yoon's (this issue) model, then, not all the branches of the genetic inheritance tree are growing simultaneously. This is another reason for considering multiple-paths or branching-tree models. The revised model also takes into accounts the tendency for local knowledge to grow into more and more general knowledge.

Nonordered learning. Constructing a really big idea can contribute to unordered or nonsequential learning. By constructing Node D1 in Figure 1, for example, students might be able to construct nodes C1, C2, and C3 in any order. (Conversely, the construction of C1, C2, or C3 could provide the basis for constructing the really big idea D1.) The same could be said for nodes B3, B4, and B5 under Node C2 and nodes A8 and A9 under node B4.

Developmental readiness (learning potential, or the zone of proximal development) and discontinuous development. Alexander et al. (1989), for example, found an example of a discontinuity in learning rates. In addition to showing that 4- and 5-year-old children benefited from explicit analogy training, the performance of some control group children also improved during the 6-month period. Those children in the control group whose performance improved were significantly older than those children who remained poor performers. Unless children are educationally and psychologically ready for a transition, their learning can be limited or delayed.

In terms of big ideas, development can be delayed until a student has had the opportunity to construct an overarching concept, even if all the subconcepts have been constructed (e.g., isolated Nodes A13 and A14 in Figure 1). Alternatively, the absence of a big idea can also rob a child of the cognitive resources needed to construct necessary subconcepts. Developmental readiness can be conceptualized as having in place the cognitive resources (nodes) needed to make a new connection.

ACKNOWLEDGMENTS

Preparation of this article was supported, in part, by grants from the National Science Foundation (BCS–0111829: "Foundations of Number and Operation Sense") and the Spencer Foundation ("Key Transitions in Preschoolers' Number and Arithmetic Development: The Psychological Foundations of Early Childhood Mathematics Education"). The opinions expressed are solely those of the author and do not necessarily reflect the position, policy, or endorsement of the National Science Foundation or the Spencer Foundation.

REFERENCES

Ainley, J., & Pratt, D. (2002). Purpose and utility in pedagogical task design. In A. Cockburn & E. Nardi (Eds.), *Proceedings of the Twenty Sixth Annual Conference of the International Group for the Psychology of Mathematics Education: Vol. 2* (pp. 17–24). Norwich, UK: UEA.

Alexander, P. A., Willson, V. L., White, C. S., Fuqua, J. D., Clark, G. D., Wilson, A. F., et al. (1989). Development of analogical reasoning in 4- and 5-year-old children. *Cognitive Development, 4,* 65–88.

Ashmann, S., Zawojewski, J., & Bowman, K. (2003). *Integrated mathematics and science teacher education courses: A modeling perspective.* Manuscript submitted for publication.

Baroody, A. J. (1984). The case of Felicia: A young child's strategies for reducing memory demands during mental addition. *Cognition and Instruction, 1,*109–116.

Baroody, A. J. (1985). Pitfalls in equating informal arithmetic procedures with specific mathematical conceptions. *Journal for Research in Mathematics Education, 16,* 233–236.

Baroody, A. J. (1987). *Children's mathematical thinking: A developmental framework for preschool, primary, and special education teachers.* New York: Teachers College Press.

Baroody, A. J. (1989a). *A guide to teaching mathematics in the primary grades.* Boston: Allyn and Bacon.

Baroody, A. J. (1989b). One point of view: Manipulatives don't come with guarantees. *Arithmetic Teacher, 37*(2), 4–5.

Baroody, A. J. (2003). The development of adaptive expertise and flexibility: The integration of conceptual and procedural knowledge. In A. J. Baroody & A. Dowker (Eds.), *The development of arithmetic concepts and skills: Constructing adaptive expertise* (pp. 1–34). Mahwah, NJ: Lawrence Erlbaum Associates, Inc.

Baroody, A. J., Benson, A. P., & Lai, M. (2003, April). *Early number and arithmetic sense: A summary of three studies.* Paper presented at the annual meeting of the Society for Research in Child Development, Tampa, FL.

Baroody, A. J., with Coslick, R. T. (1998). *Fostering children's mathematical power: An investigative approach to K-8 mathematics instruction.* Mahwah, NJ: Lawrence Erlbaum Associates, Inc.

Baroody, A. J., & Lai, M. (2002, April). *Defining the whole when solving fraction problems.* Paper presented at the annual meeting of the American Educational Research Association, New Orleans, LA.

Baxter, J. A., Woodward, J., & Olson, D. (2001). Effects of reform-based mathematics instruction on low achievers in five third-grade classrooms. *Elementary School Journal, 101,* 529–547.

Beishuizen, M. (1993). Mental strategies and materials or models for addition and subtraction up to 100 in Dutch second grades. *Journal for Research in Mathematics Education, 24,* 294–323.

Boaler, J. (1998). Open and closed mathematics: Student experiences and understandings. *Journal for Research in Mathematics Education, 29,* 41–62.

Bottge, B. A., Heinrichs, M., Chan, S., & Serlin, R. C. (2001). Anchoring adolescents' understanding of math concepts in rich problem solving environments. *Remedial and Special Education, 22,* 299–314.

Bottge, B. A., Heinrichs, M., Mehta, Z., & Hung, Y. (2002). Weighing the benefits of anchored math instruction for students with disabilities in general education classes. *The Journal of Special Education, 35,* 186–200.

Bruner, J. S. (1966). *Toward a theory of instruction.* Cambridge, MA: Harvard University Press.

Clements, D. H., & McMillen, S. (1996). Rethinking "concrete" manipulatives. *Teaching Children Mathematics, 2,* 270–279.

Cobb, P., Wood, T., & Yackel, E. (1991). A constructivist approach to second grade mathematics. In E. von Glasersfeld (Ed.), *Constructivism in mathematics education* (pp. 157–176). Boston: Kluwer.

Dewey, J. (1963). *Experience and education.* New York: Collier.

Fuson, K. C., & Burghardt, B. H. (2003). Multidigit addition and subtraction methods invented in small groups and teacher support of problem solving and reflection. In A. J. Baroody & A. Dowker (Eds.),

The development of arithmetic concepts and skills: Constructing adaptive expertise (pp. 267–304). Mahwah, NJ: Lawrence Erlbaum Associates, Inc.

Fuson, K., Carroll, W., & Drueck, J. (2000). Achievement results for second and third graders using the Standards-based curriculum *Everyday Mathematics*. *Journal for Research in Mathematics Education, 31,* 277–295.

Gagné, R. M., & Briggs, L. J. (1974). *Principles of instructional design.* New York: Holt, Rinehart & Winston.

Ginsburg, H. P. (1977). *Children's arithmetic.* New York: Van Nostrand.

Ginsburg, H. P., & Baroody, A. J. (2003). *Test of early mathematics ability* (3rd ed.). Austin, TX: Pro-Ed. (Original work published in 1983)

Ginsburg, H. P., Klein, A., & Starkey, P. (1998). The development of children's mathematical knowledge: Connecting research with practice. In I. E. Sigel & K. A. Renninger (Eds.), *Handbook of child psychology: Vol. 4. Child psychology in practice* (5th ed., pp. 401–476). New York: Wiley.

Gravemeijer, K. (2002, April). *Learning trajectories and local instruction theories as a means of support for teachers in reform mathematics education.* Paper presented at the annual meeting of the American Educational Research Association, Las Vegas, Nevada.

Griffin, S. A., Case, R., & Siegler, R. S. (1994). Rightstart: Providing the central conceptual prerequisites for the first formal learning of arithmetic to students at risk for school failure. In K. McGilly (Ed.), *Classroom lessons: Integrating cognitive theory and classroom practice* (pp. 25–49). Cambridge, MA: MIT Press.

Hatano, G. (1988). Social and motivational bases for mathematical understanding. In G. B. Saxe & M. Gearhart (Eds.), *Children's mathematics* (pp. 55–70). San Francisco: Jossey-Bass.

Hatano, G. (2003). Forward. In A. J. Baroody & A. Dowker (Eds.), *The development of arithmetic concepts and skills: Constructing adaptive expertise* (pp. xi–xiii). Mahwah, NJ: Lawrence Erlbaum Associates, Inc.

Hiebert, J., & Lefevre, P. (1986). Conceptual and procedural knowledge in mathematics: An introductory analysis. In J. Hiebert (Ed.), *Conceptual and procedural knowledge: The case of mathematics* (pp. 1–27). Hillsdale, NJ: Lawrence Erlbaum Associates, Inc.

Kuhn, T. S. (1970). *The structure of scientific revolutions.* Chicago: University of Chicago Press.

Mix, K. S., Huttenlocher, J., & Levine, S. C. (2002). *Math without words: Quantitative development in infancy and early childhood.* New York: Oxford University Press.

Miura, I. T., & Okamoto, Y. (2003). Language supports for mathematics understanding and performance. In A. J. Baroody & A. Dowker (Eds.), *The development of arithmetic concepts and skills: Constructing adaptive expertise* (pp. 229–242). Mahwah, NJ: Lawrence Erlbaum Associates, Inc.

National Council of Teachers of Mathematics. (1989). *Curriculum and evaluation standards for school mathematics.* Reston, VA: Author.

National Council of Teachers of Mathematics. (1991). *Professional standards for teaching mathematics.* Reston, VA: Author.

National Council of Teachers of Mathematics. (1995). *Assessment standards for school mathematics.* Reston, VA: Author.

National Council of Teachers of Mathematics. (2000). *Principles and standards for school mathematics: Standards 2000.* Reston, VA: Author.

Paulos, J. A. (1991). *Beyond numeracy: Ruminations of a numbers man.* New York: Knopf.

Piaget, J. (1964). Development and learning. In R. E. Ripple & V. N. Rockcastle (Eds.), *Piaget rediscovered* (pp. 7–20). Ithaca, NY: Cornell University.

Piaget, J., & Beth, E. (1966). *Mathematical epistemology and psychology.* Dordrecht, The Netherlands: Reidel.

Resnick, L. B. (1982). Syntax and semantics in learning to subtract. In T. P. Carpenter, J. M. Moser, & T A. Romberg (Eds.), *Addition and subtraction: A cognitive perspective* (pp. 136–155). Hillsdale, NJ: Lawrence Erlbaum Associates, Inc.

Resnick, L. B., & Ford, W. W. (1981). *The psychology of mathematics for instruction.* Hillsdale, NJ: Lawrence Erlbaum Associates, Inc.

Sagan, C. (1997). *The demon-haunted world: Science as a candle in the dark.* New York: Ballantine.

Seo, K., & Ginsburg, H. P. (2003). "You've got to carefully read the math sentence …": Classroom context and children's interpretations of the equals sign. In A. J. Baroody & A. Dowker (Eds.), *The development of arithmetic concepts and skills: Constructing adaptive expertise* (pp. 161–187). Mahwah, NJ: Lawrence Erlbaum Associates, Inc.

Sarama, J., Clements, D. H., & Vukelic, E. B. (1996). The role of a computer manipulative in fostering specific psychological/mathematical processes. In E. Jakubowski, D. Watkins, & H. Biske (Eds.), *Proceedings of the Eighteenth Annual Meeting of the North America Chapter of the International Group for the Psychology of Mathematics Education* (Vol. 2, pp. 567–572).

Simon, M. (1995). Reconstructing mathematics pedagogy from a constructivist perspective. *Journal for Research in Mathematics Education, 26,* 114–145.

Sowder, J., Philipp, R., Armstrong, B., & Schappelle, B. (1998). *Middle-grade teachers' mathematical knowledge and its relationship to instruction.* Albany, NY: State University of New York.

Spelke, E. (2003a, April). *What makes humans smart?* Paper presented at the biennial meeting of the Society for Research in Child Development, Tampa, FL.

Spelke, E. (2003b). What makes us smart? Core knowledge and natural language. In D. Genter & S. Goldin-Meadow (Eds.), *Language in mind.* Cambridge, MA: MIT Press.

Starkey, P., & Cooper, R. G. (1995). The development of subitizing in young children. *British Journal of Developmental Psychology, 13,* 399–420.

Steier, F. (1995). From universing to conversing: An ecological constructionist approach to learning and multiple description. In L. P. Steffe & J. Gale (Eds.), *Constructivism in education* (pp. 67–84). Hillsdale, NJ: Lawrence Erlbaum Associates, Inc.

Thorndike, E. L. (1922). *The psychology of arithmetic.* New York: Macmillan.

van Hiele, P. M. (1986). *Structure and insight.* Orlando, FL: Academic Press.

von Glasersfeld, E. (1982). Subitizing: The role of figural patterns in the development of numerical concepts. *Archives de Psychologie, 50,* 191–218.

Vygotsky, L. S. (1978). *Mind in society: The development of higher psychological processes.* Cambridge, MA: Harvard University Press.

Woodward, J., & Baxter, J. (1997). The effects of an innovative approach to mathematics on academically low achieving students in inclusive settings. *Exceptional Children, 63,* 373–388.

Wynroth, L. (1986). *Wynroth math program—The natural numbers sequence.* Ithaca, NY: Wynroth Math Program. (Original work published in 1975)

Contributor Information

Mathematical Thinking and Learning, a fully refereed journal, is directed at researchers interested in mathematics education from the perspective of psychology, sociology, philosophy, anthropology, mathematics, and information technology, with a particular focus on mathematical thinking, reasoning, and learning.

The journal seeks original, high-quality articles that address one or more of the following topics: interdisciplinary studies on mathematical learning, reasoning, and thinking, and their developments at all ages; technological advances and their impact on mathematical thinking and learning; studies that explore the diverse processes of mathematical reasoning; new insights into how mathematical understandings develop across the life span, including significant transitional periods; changing perspectives on the nature of mathematics and their impact on mathematical thinking and learning in both formal and informal contexts; studies that explore the internationalization of mathematics education, together with other cross-cultural studies of mathematical thinking and learning; and studies of innovative instructional practices that foster mathematical learning, thinking, and development.

In addition to receiving research articles, the journal invites articles that present theoretical and philosophical analyses of issues related to the previous topics.

Submission: Submit four copies and a PC-formatted disk copy of the manuscript to the Editor, Lyn D. English, Centre for Mathematics and Science Education, Queensland University of Technology, Victoria Park Road, Kelvin Grove, Brisbane, Australia 4059 (e-mail: L.English@qut.edu.au).

Cover Letter: The cover letter should include the contact author's complete mailing address, e-mail address, and telephone and fax numbers. In the cover letter, the author(s) should request publication of the manuscript in *Mathematical Thinking and Learning* and should include (a) a statement that the manuscript is not previously published or simultaneously submitted elsewhere (manuscripts copyrighted electronically or online will also not be considered); (b) confirmation that all original research procedures were consistent with the principles of the research ethics published by the American Psychological Association, except as may be detailed in the manuscript; and (c) a request for blind review, if desired.

Manuscript Preparation: Manuscripts should be prepared according to the guidelines of the *Publication Manual of the American Psychological Association* (5th ed.). Double space all material and place in the following order: title page, abstract, text, quotations, acknowledgments, references, appendixes, footnotes, tables, and figures. The title page should include the title of the manuscript; names and affiliations of all authors; name, mailing address, e-mail address, and telephone and fax numbers of the corresponding author; and a running head of no more than 48 letters and spaces. The second page should include the manuscript title and an abstract of no more than 100 to 150 words. Number all manuscript pages, including the figures. All figures must be camera-ready black-and-white originals.

Permissions: Authors are responsible for all statements made in their work and for obtaining permission from copyright owners to use a lengthy quotation (500 words or more) or to reprint or adapt a table or figure published elsewhere. Authors should write to original author(s) and publisher of such material to request nonexclusive world rights in all languages for use in the article and in future editions. Provide copies of all permissions and credit lines obtained.

Production Notes: Authors are sent page proofs of their manuscripts and are asked to proofread them for printer's error and other defects. Correction of typographical errors and query responses will be made free of charge; other alterations will be charged to the author. Authors may order reprints of their articles when they receive page proofs. Printing considerations do not permit the ordering of reprints after authors have returned proofs.

.